# POWERING COLONIALISM

INTERSECTIONS:
HISTORIES OF ENVIRONMENT,
SCIENCE, AND TECHNOLOGY
IN THE ANTHROPOCENE

SARAH ELKIND AND FINN ARNE JØRGENSEN, EDITORS

# POWERING COLONIALISM

## ELECTRIFICATION, EXTRACTION, AND EMPIRE IN AOTEAROA NEW ZEALAND

NATHAN N. KAPOOR

UNIVERSITY OF PITTSBURGH PRESS

Published by the University of Pittsburgh Press, Pittsburgh, Pa., 15260


Printed on acid-free paper

10 9 8 7 6 5 4 3 2 1

Cataloging-in-Publication data is available from the Library of Congress

Hardcover: 978-0-8229-4887-2
Paperback: 978-0-8229-6813-9

Cover art: *Dominion Day at the General Assembly Library Building,* William Hall Raine.
Cover design by Melissa Dias-Mandoly

Map figures produced in collaboration with John Kostelnick, professor of geography at Illinois State University.

Publisher: University of Pittsburgh Press, 7500 Thomas Blvd., 4th floor, Pittsburgh, PA 15260, United States, www.upittpress.org
EU Authorized Representative: Easy Access System Europe, Mustamäe tee 50, 10621 Tallinn, Estonia, gpsr.requests@easproject.com

# CONTENTS

Note on Māori Language
vii

Preface: "Energy Creek"
xi

Acknowledgments
xv

INTRODUCTION:
Electrification as New Zealand's Colonial Imaginary
3

CHAPTER 1.
Reefton: New Zealand's Electrical Utopia
25

CHAPTER 2.
Mining for Power: Electrifying the Phoenix Mine and the Creation of a Model Colony
54

CHAPTER 3.
The Station at World's End:
Settler Colonialism in the Lake Coleridge Scheme
76

CHAPTER 4.
The Megawattasaurus: The Huntly Power Station
and New Zealand's Energy Regime
103

CHAPTER 5.
"A Pebble in Your Shoe":
Geothermal Power and Settler Colonialism
137

CONCLUSION:
Electrification Is Not a Finished Process
167

Appendix: Timelines for the General History of Aotearoa
New Zealand and Its Electric Power Bureaucracy
175

Notes
177

Bibliography
219

Index
243

# NOTE ON MĀORI LANGUAGE

Although I am not fluent in the Māori language, throughout this book I use Te Reo Māori to talk about people, places, traditions, ideologies, and culture because the language is integral to the history and ongoing conversations about settler-colonialism in Aotearoa New Zealand. Scholars who study New Zealand may find the need for such notice unnecessary because most recent scholarship on the people of New Zealand incorporates this language. However, as I expect a sizable community of scholars who might work with this text will be unfamiliar with the language and the history of its use, I hope this explanation will help establish a stronger connection.

As a novice and learner, I acknowledge that my interpretations and use of Te Reo Māori terms could be conveyed in other ways. I do not seek to carelessly insert Indigenous language into this history, as many have noted about public signage and education reform in New Zealand.[1] Instead, I use it to insist upon the necessity of the Māori in this history. While Te Reo Māori's history is tied to settler-colonialism because of the way invaders classified it, it is also a language of strength that Māori revitalized throughout the twentieth century.

Right now, New Zealanders debate even the country's name. Should it be New Zealand, Aotearoa (Māori for "land of the long white cloud"), or Aotearoa New Zealand?[2] Not all settlers (non-Indigenous New Zealanders) and not all Māori agree on deploying Aotearoa as a name for the country, which is why I use the term at intervals throughout the book. Te Reo Māori's place in Aotearoa New Zealand remains contested, but the language is alive.

Therefore, in step with some other historians of the Māori and New Zealand, Māori words are not italicized unless the author of a cited source did so. Although italicization can simply signal a foreign language or provide useful contrast for the reader on a new term, it also can unintentionally marginalize.[3] Te Reo Māori is not dead. The Māori are not "conquered" or trapped in the past. At the same time, the language is not universal. There are phrases and words that some Māori communities assign different meanings. Sometimes English and Te Reo Māori flow back and forth, but unlike the public signage in New Zealand, which usually features both languages, sometimes translation is not at all straightforward. As Alison Jones beautifully explores in *This Pakeha Life*, *whare* might be translated as "hut" or "house" or "a place to encounter spirits." She notes, "There is something about both that shift of vowel sound and the caution at the imagined whare door that captures the disjuncture between Pakeha and Māori worlds."[4]

My deployment of Māori words comes from expert translations, Māori speakers, and my use of John C. Moorfield's *Te Whanake Te Aka: Māori-English Dictionary* to support my translations.[5] Apart from the words I use, I want to draw attention to one word I do not use—*Pākehā*. It is a term burdened by a history I do not thoroughly explore, and how New Zealanders understand it remains in flux. Although some equivalents exist in other settler colonial spaces, the contestation surrounding what it means to be Pākehā is distinct. As Ani Mikaere argued, "There is nowhere else in the world that one can be Pākehā. Whether the term remains forever linked to the shameful role of the oppressor or whether it can become a positive source of identity is up to Pākehā themselves. All that is required from them is a leap of faith."[6] Although some historians wield *Pākehā* effectively, and the term can expose the history of settler-colonialism in New Zealand, I do not deploy it in this book because I rely on the term *settler* to link New Zealand's story with other colonial stories worldwide while writing as an outsider.

Past and present peoples of Aotearoa New Zealand do not often fit neatly into the categories of Māori or Pākehā. Indeed, many scholars and activists have explored how we might best describe them and tell their stories without those terms. This issue brings me to another term that I do not deploy, but is nonetheless useful to consider: Tangata Tiriti, or "people of the treaty." Avril Bell's recent work, *Becoming Tangata Tiriti: Working with Māori, Honouring the Treaty,* demonstrates that beyond the myriad complexities that exist between the categories of Māori and Pākehā, we might deploy other categories to describe the relationships and partnerships between these groups.[7] While I did not adopt that language here, I have worked to make my usage of *Māori* and *settler* an active part of the conversation about how we understand the people of New Zealand and their relationship to their colonial past. I chose these terms because I think they fit best within the history of electrification of New Zealand, not because they are authoritative or absolute. By understanding their fluidity, controversy, and utility, historians are better suited to listen to the stories the people tell rather than box them into a particular narrative.

FIGURE P.1. Dry Creek (i.e., Energy Creek), December 22, 2023. Photo by author.

# PREFACE

# "ENERGY CREEK"

About a mile from the house where I grew up in East Tennessee, a small creek, locally known as Dry Creek, ran alongside the route to school. The stream is the product of local springs, runoff from Little Mountain off Laurels Road, and watershed further uphill from Watauga Lake. *Watauga* is a Cherokee word that marks towns and bodies of water from western North Carolina to Texas. Like so much Cherokee land, culture, resources, and history, settlers appropriated the term to name and define their property. Public works engineers created the lake during the early twentieth century to harness the region's water resources for generating electric power, culminating in the Tennessee Valley Authority's construction of Watauga Dam, a significant source of the area's electricity. Dry Creek is a part of this history. However, my parents tell me that, as a kid, without any interest or awareness of the politics of hydropower, I came up with a different name for the creek: "Energy Creek."

Of course, I can never precisely explain my naming the creek, but I remember thinking about the white water's speed, violence, and power. The creek changed drastically depending on the amount of rain we received and how many rocks my friends and I piled up to redirect the flow. During the summer months, the water remained barely a foot deep and trickled through the forest so that you could hear it while driving on Dry Creek Road with the window down—I still do this when I return home. However, when it rained, the creek, like so many rivers in the region, turned a light brown (because of the massive quantity of clay in the soil), then raged, becoming a small but turbulent whitewater rapid. That is what I remember the most. I can see the channels of white water flailing on the water's surface, cracking against the rocks, and once even taking out a small driveway bridge. I remember being obsessed with it and wanting my parents to stop and let me watch the water churn.

My understanding of water as energy, even as a small child, is telling. But why did I call it Energy Creek?

"Energy Creek" and the Inangahua River, the watercourse that drove the first New Zealand hydropower station I explore in chapter 1 of this book, are approximately 8,573 miles apart. Despite residing on different landmasses, they share many qualities. Both sit on land that white-settler governments, public works infrastructures, and extractive industries utilized while stealing and profiting from.[1] Both are part of hydroelectric systems. Both are fed by mountain runoff. And I have analyzed and obsessed over both bodies of water. I am a white American from East Tennessee and a quarter Indian with the only connection to my family there being in name alone. I grew up and received my secondary and undergraduate education in Tennessee. I am a product of a settler society, and when I came to study the history of power systems in New Zealand, I approached the archive as a settler. My primary archives were those of the colonizers and settlers, even as I increasingly addressed histories and stories of Māori, the Indigenous peoples of New Zealand. To critique the colonial foundations of New Zealand's electric power infrastructure, I must situate myself with the places, peoples, and systems I include.

Although my labeling of "Energy Creek" may have been an innocuous turn of phrase at the time, it nonetheless encapsulated settler rhetoric and ideology: Nature is to be exploited for energy in service to production and self-sufficiency. And for all my attempts in this history to wield theoretical tools and historical narratives that expose that mentality, I undoubtedly succumbed to that logic in my understanding the development of electric power systems in New Zealand.

At the beginning of my research, when I first discovered mention of New Zealand's hydroelectric networks at a gold mine, it would have never occurred to me to think critically about their association with "Energy Creek"—I probably would have dismissed it. Now, they are inseparable in my mind. I chose to study New Zealand because the electrical engineer's settler rhetoric was foundational to how I thought about the relationship between energy and natural resources. I did not recognize it as electric infrastructure bound to colonialism. What follows is my attempt to build on the scholarship and experiences of those inside and outside of New Zealand who lived with, experienced, and pushed back against colonial power systems. In New Zealand, as in East Tennessee, energy and electric power were and are shaped by settler language, policy, and actions. We must expose this reality and acknowledge our place in it.

As a settler who grew up near Energy Creek, I had to learn (and am still learning) to see colonial power—many do not have that privilege. In attempting transparency in my scholarship, I aim to give readers the tools to understand that scholarship. I want this transparency to challenge my work, and I embrace the possibility that it might dissuade readers from engaging. Last, I hope these words continually remind me of the potential utility and limits of my exploration of colonialism and electrification in New Zealand.

# ACKNOWLEDGMENTS

I do not remember who said it to me, but at some point in school, someone told me that writing a dissertation is a lesson in how not to write one. As I worked on this book, I took comfort in knowing I was learning how not to write a book. One reason I do not remember who gave me this insight is that I have been incredibly fortunate to have been surrounded by so many supportive people throughout my career. Perhaps it is trite, but I cannot possibly thank everyone; nonetheless, I am going to try. After all, this is probably just another lesson in how not to write acknowledgments.

I am indebted to the institutional and financial support of the History of Science Department and the Graduate College at the University of Oklahoma. They helped me facilitate travel and extended stays in Wellington and Christchurch to access archives. Following my time there, Grand Valley State University and Illinois State University also provided resources, writing workshops, and travel assistance. I am especially grateful to the Office of Research and Graduate Studies at Illinois State University for their support in the final steps of the publication process. During my time working in Aotearoa New Zealand, I received so much guidance from archivists and staff at the Archives

New Zealand and the Alexander Turnbull National Library. They often helped me find what I did not know I needed. Furthermore, in communication with the Milner Library staff at Illinois State University, they often helped me acquire digital materials from archives and libraries around Aotearoa New Zealand when travel was not possible.

Within and beyond each of these institutions, I am indebted to so many people. To George Webb at Tennessee Technological University for being the first to teach me about and inspire me to pursue the history of science and technology and for continuing to be a critical and supportive mentor; to Anna Reser and Leila McNeill for those post-seminar sessions at the Library, and for showing me what it meant to be a good colleague and pushing me to be better; to Suzanne Moon, Hunter Heyck, Pete Soppelsa, Bill Storey, Robert Lifset, Peter Barker, and Piers Hale for their work and support during and after my time in graduate school; to the other contributors to the edited volume *Electrical Conquest* who helped me situate my work alongside recent histories of electrification, especially Abby Spinak for her thoughtful critique and discussions that altered how I approached this book; to the Energy History Working Group, especially Robert Lifset, Sarah Stanford-McIntyre, and Chris Jones, who read early versions of several chapters.

While traveling during my initial research, I had the privilege of being hosted by incredibly gracious hosts and communities. Thank you to Maya, Jason, and Nath for hosting me, cooking for me, and giving me a temporary home on the other side of the world. Not only did I have a comfortable place to stay; you also introduced me to your cities and communities in ways that shaped how I saw this project. I also appreciate the time and transportation offered by William and Nicole. Without you, I would not have been able to see Lake Coleridge and explore Christchurch. Thanks for encouraging me to explore and for making my time outside the archives so much less isolating.

At Illinois State University, I have been fortunate to find colleagues who transformed the work into community. I am grateful to Christine Varga-Harris and Katie Jasper for their mentorship and willingness to tolerate my random advice-seeking visits. I am especially thankful for Keith Pluymers, who has been something like an older brother in the best possible sense. From the day I arrived on campus, he has been someone who took me seriously and who has consistently pushed me in directions I would not have found on my own. His willingness to challenge my assumptions has always come wrapped in the kind of care that makes the challenge feel like a gift. To Camille Cole, whose strong and genuinely thoughtful questions pushed me to reckon with the arguments I was making rather than the ones I wished I was making—and whose generosity in sharing her time and resources reminded me, more than once, that good colleagues make you better. To Lindsay Stallones-Marshall, for being exactly the kind of teacher-scholar I hope to be: one who thinks as seriously about the craft of teaching as the craft of writing, and with whom I could talk about both without having to explain why either mattered. And finally, to Taylor Soja, who was hired alongside me and who has somehow made working alone on a book a collaborative project. Taylor has a remarkable and almost unsettling ability to understand what I am trying to say before I have found the words. She has also seen me through insecurities and the self-doubt and has met all of it with a generosity that I will never be able to fully repay. There is no adequate way to account for what it means to find your people, but I am grateful, every day, that I did.

No acknowledgment of any of my scholarship will ever be complete without recognizing Suzanne Moon, whose influence runs through every chapter whether or not her name appears in the footnotes. I can never repay the reading lists, hours of meetings, and draft markups. As my doctoral advisor, she modeled what it means to take ideas seriously without taking yourself too seriously, to push a student toward their best thinking while trusting them to find their own way there. It took me years to realize how masterfully she balanced guiding my research with letting me try and, sometimes, fail while always being there to pick me back up. She showed me how to get lost in questions and to embrace frustration with archives and writing. Whatever clarity this book achieves, I owe to her.

I am also grateful to the people at the University of Pittsburgh Press who shepherded this project from a collection of chapters into an actual book. Abby Collier and Amy Sherman guided me through peer reviews and multiple rounds of revision with a patience I did not always deserve. I am, by nature, an anxious author, and they met every uncertain email and worried question with nothing but encouragement and good sense. A first book is an act of faith, and it helped enormously to have people in my corner who seemed to believe in it even when I was not sure I did.

I owe a different kind of debt to my family, one that long predates this project or any archive. My mother gave me a love of stories before I knew that stories could be history. Not just because she loved our local storytelling festival, which she very much does, but because of the stories she shared of her work as a nurse and about her patients. Finally understanding in recent years that my mother inspired me to pursue history has been a quiet constant. My father taught me, by example, that in life—and, as it turns out, in historical scholarship—balancing work and wonder is key. He practices and preaches that learning is not something that ends, a fact I am only just beginning to appreciate about his guiding hands over the years. In what I expect may be a surprise to her, because we are (for better or worse) siblings of so few words, I want to thank my sister, Sarah, who showed me what perseverance actually looks like. She's stood up to challenge after challenge and never once quit. I carry something from each of them into everything I write and the work I do.

And then there are the humbling debts. My daughters, Marie and Cora, have been perhaps my most recent and effective teachers, daily and without mercy reminding me of the vast distance between what I think I know and what there is still to learn. As for Amber, she has only ever steadied me when the project felt impossible, and has, in every way that matters, only ever made me better. This book is for her too.

# POWERING COLONIALISM

## INTRODUCTION

# ELECTRIFICATION AS NEW ZEALAND'S COLONIAL IMAGINARY

In a press release for the Labour Party in 2019, Aotearoa New Zealand Prime Minister Jacinda Ardern said, "We are committed to taking the next steps to get to 100 percent renewable with support for projects like pumped hydro at Lake Onslow, removing the barriers to development of new renewable electricity projects, advancing green hydrogen and other green technologies and supporting businesses to decarbonise. Our plan for clean energy and lower carbon emissions will help us seize the economic opportunities of being the clean, green country that New Zealanders see ourselves as being and that we can market ourselves as."[1] As suggested by some of Ardern's claims, electric power issues occupy people's attention in New Zealand, including grid security after Cyclone Gabrielle in 2023 and concerns about the fate of New Zealand's Zero Carbon Act 2019. The company Emrod, founded in 2019, plans to kickstart a global shift toward wireless electric power technology.[2] Beyond the work of firms like Emrod, New Zealand's electric power initiatives and projects feature in electrification programs around the world. For example, Māori-led geothermal power initiatives in Kenya and the United States have initiated global conversations about how Indigenous peoples might use geothermal sources to provide electricity and foster socioeconomic relief.[3] In short, New Zealand's electric power infrastructure looks poised to deliver exciting systems and outcomes inside and outside the country in the coming decades.

New electric power technologies inspire interest as technological quick fixes toward a solution to systemic energy inequity. Partly this is because they offer to expand a particular vision in which electricity allows for the production and consumption of more. Such contemporary plans build on a long history of electric power schemes in New Zealand, all of which have promised more—more production and more consumption for *some* people. Beginning in the 1880s, settler New Zealanders developed hydroelectric systems to light towns, drive extractive industries, communicate, and integrate technologies the British Empire deemed modern into the colony. Throughout the twentieth century, the state diversified the energy sources used to generate electric power by adding fossil fuels and geothermal energy into its electric infrastructure. The development of these systems mirrors electrification stories from around the world. What distinguishes New Zealand's story is how the people that designed and used electric power systems saw them as the best vehicles for achieving settler visions for the country. Electrification in New Zealand is a form of settler-colonial logic espoused by boosters and public works authorities in New Zealand and worldwide. New Zealand's electrification story demonstrates how the global history of these initiatives is rooted in, and profits from, extractive and inequitable practices engrained in colonial politics and ideologies. Building on significant additions to energy history from environmental, cultural, technology, and political historians, I argue that electrification must be understood as settler-colonialism within the larger transition to state control of energy resources, use of fossil fuels, and massive industrial production models.

Framing electrification as part of the settler-colonial project in Aotearoa New Zealand clarifies how and why electrification unfolded there. The availability of technology alone did not guarantee its adoption. So, when we consider things like Emrod's plans for wireless electric power—seen as innovative in 2023, even though such ideas circulated among engineers over a century earlier—historians are better positioned to explain how certain electric systems came to dominate. Many in the engineering and scientific community are quick to argue that technological, scientific, and material limits to electrification exist; therefore, electric power systems *had to* develop in a specific way. Furthermore, historians of electric power technology note that there are limits to sociocultural acceptance. The claims from both groups are, at most, partially true. There are indeed material limits to producing and harnessing electric power, just as people embed some technologies into their worldview while finding others incompatible. This book is about how such ideas developed in New Zealand, and why people choose or believe in certain energy systems.

From its earliest electrified mines to Ardern's speeches about energy futures, New Zealand's power production and consumption have been closely associated with labels like *self-sufficient, sustainable,* and *green*. Following how the country electrified provides a way to explore how these labels and understandings of energy use took shape. If indeed Aotearoa New Zealanders—indeed, much of the world—see the country's electrification as a model energy identity, then the country's energy infrastructure must be clearly linked to the sociopolitical systems that created those labels. While terms like *self-sufficiency, sustainable,* or *green* might typically be seen as aspirational goals linked to contemporary environmental movements, energy efficiency prerogatives, or broad initiatives to combat anthropogenic climate change, in New Zealand they are rooted in the settler-colonial practices that used electrification as a tool to effect the goals of the British Empire, the colonial government, and settlers.

As historians like Fredrik Meiton (*Electrical Palestine*) or Andrew Needham (*Power Lines*) argue, one of the issues at stake when asking about how people electrified is, why did people begin to use specific energy resources—hydro, mining, geothermal—for electrification?[4] The question remains worth asking, whether during the earliest days of speculation about the wonders of electric currents, or today, when countries worldwide grapple with how the continued production of electric power can either help destroy or save our planet. In New Zealand, since at least 1886, electric power systems facilitated settler-colonial imaginaries and realities that settlers and Māori built into New Zealand's electric power infrastructures. They continue to inform how people think about energy. The history of New Zealand's electrification reveals how electric power technologies dramatically developed, thrived, and transformed around settler-colonial ideas about the purpose of the islands' resources and peoples.

In recent years, discussion of electric power systems' sociopolitical impact has flourished in mainstream and scholarly literature, stimulated by such energy-related concerns as global warming, overpopulation, and increasing energy inequality. Histories of electrical science and technology have been popular for decades. Recently, historians, sociologists, political scientists, and anthropologists have felt the need to respond to energy crises in their work—a conversation I aim to influence here.[5] Writing about the history of electricity while plugging in our nearest devices—or reading such histories on them—we are complicit and intertwined with those infrastructures. It can be difficult to imagine escaping them or even wanting to do so. However, accepting electric power as inevitable is to comply with settler-colonial logic. I choose not to treat electrification as inevitable, and I encourage other historians and users of electricity to be critical of the choice to use electricity—in the past and present. Interrogating our electrical past is the only way to appreciate how electrical systems shape our world and might be altered to better navigate climate change and inequality.

## ENFORCING ELECTRIFICATION AND EXPLOITATION

In many histories of electrification, particularly the majority written about New Zealand, the complexity and scale of electrical systems have eclipsed the salience of their ideological and cultural influence. As much as New Zealand's transition to electric power reflects the actions for and against the formation of a state, it is also about the cultural significance of electricity to British politicians, European settlers, and the Māori, as well as the use of technology to enforce, ignore, and resist colonialism. Electrification is part of a seamless web of connections among people, machines, and the environment. Electric systems must account for the input of politicians, scientists, administrators, legal operatives, engineers, entrepreneurs, and businesspeople, because all of these perspectives respond to technical problems and are involved in developing electrification. Yet the resistance to electric power or its associated infrastructure is just as noteworthy in shaping electricity usage as the networks that build and regulate it. And perhaps more importantly, electricity itself is not a neutral energy source waiting to be used at the right time. It was an active and politically charged element of British society that made it attractive to certain classes and unattractive to others for reasons that had little to do with its users' actual capabilities. For many colonial spaces in the British Empire, technocrats used electric power to delineate between "white" and "dark," reinforcing racial hegemony. Electric power and electrical technologies in this book demonstrate that energy shifts must be dealt with not only at the labs, drafting tables, or boardrooms but also at the propagandist fundraisers, town council meetings, and protests. Further, the relevant people and places involved may not easily be thought of as coexisting under the auspices of the state.

Understanding the sociopolitical impact of transitions to electricity is not an unprecedented move in the history of technology or electricity, of course.[6] Since the 1980s and, more formally, from the 1990s on, historians of electricity have offered narratives of electrification that treat electrical systems as massive social and technical networks. What is new is that they highlight how New Zealand's electrification within larger energy transitions positions researchers to challenge the assumption that electrification was inevitable throughout the world during the late nineteenth century and into the present.[7] The success of electrification did not hinge solely upon the successful mastery of electromagnetic theory or refinement of conductive materials. It is both a product and cause of a broader reconceptualization of energy that took place across the world throughout the nineteenth and twentieth centuries. Understanding electrification, in the British Empire, New Zealand, and beyond as a foreordained outcome of industrialization ignores the role colonials' assumptions played in the change of energy systems. The process of electrification in New Zealand, therefore, cannot be understood without unpacking the colonial machinations embedded in the grid that settlers and Māori navigated. The connection of New Zealand and its electrification to settler-colonialism will expand both how historians see those networks and how historians of New Zealand specifically understand settler-colonialism.

To appreciate why colonialism played a formative role in New Zealand's electrification and to avoid deterministic assumptions about electrification, I approach the country's electrification as part of a much longer energy transition. An *energy transition* refers to a structural change in energy systems, including a shift in energy resources, distribution, and use. These transitions take place over prolonged periods in which people, governments, and companies seek more satisfactory means of using energy for power, light, and transportation.[8] Situating New Zealand, and indeed many other states' electrification narratives in the 1880s, within these transitions is necessary for helping us better trace the sociopolitical forces that governed the course of electric power technologies. The scale of the transition expands the timeline, geographic scope, and number of power sources involved in electrification. While mines, cities, and companies built electric power systems at the local level, the shift represents global transitions in how people harvest, store, produce, and consume energy. Even though this study focuses on New Zealand, I treat its electrification as part of a global energy transition.

Energy systems are complex sociotechnological arrays that involve the technologies associated with production and the humans that design and make the technologies, manage energy distribution, and use and consume energy. Therefore, efforts to change energy systems must account for the broader social and economic assemblages built around energy production and consumption. Framing shifts in energy systems as part of a more extended transition disrupts ideas about the stability or predetermination of power sources by locating the sociocultural influences that motivated the search for new energy systems. The consequences of an energy transition go beyond the economic costs or the exciting implications of new energy technologies, and so must our historical investigations. New energy systems have immediate and far-reaching sociopolitical consequences that redefine how people live, work, and govern.[9] Recognizing these consequences shows how energy resources and systems have benefited oppressive political regimes and how those same systems can be used to challenge oppression.[10] Environmental and energy historians have done much to show how resources—particularly energy resources—are bound to the violent dispossession wrought by colonialism.[11] Andrew Curley argues that the word *resources* is just a synonym for colonialism.[12] The extraction of resources and the expansion of state-run infrastructure cannot be disentangled from the settler-colonial prerogative. Because electric power technology served as a tool for extracting, controlling, and organizing energy, mineral, agricultural, and transportation resources, it was colonial from the start.

Electrical systems grew rapidly at the turn of the twentieth century, and a broad literature depicts how their infrastructures expanded, consolidated, and at times stagnated.[13] These power networks flourished through an amalgamation of governmental, private, and corporate funding; collective innovation; and the birth of electrical engineering. Furthermore, these systems spanned massive geographic spaces. New Zealand's stories of remote lighting of towns and mines are connected to the simultaneous construction of large electrical systems elsewhere, especially throughout the British Empire. This book expands the timeline typically associated with the implementation of electric power systems in New Zealand and gives more attention to the global scope of electrification. Allowing for a longer timespan better accounts for the sociopolitical motivations behind a transition that predates dreams of centralized power production or grids. For instance, I tie early-nineteenth-century colonial patterns of resource usage to the rhetoric of electrification boosters during the late nineteenth century. New Zealand's electrification relied on settler and Māori use of water and geothermal resources, which had profound implications for New Zealand's history, such as the massive growth on the North Island, the expansion of agriculture around Christchurch, and the growth of the mining industry in Otago.

As David Nye argues in *Electrifying America,* and more recently in *American Illuminations,* electrification was not a "'natural' or 'neutral' process; everywhere, it was shaped by complex political, technical, ideological interaction."[14] Appreciating the British "culture of energy" is crucial for understanding the ready reception of, and at times skepticism toward, electric power systems in Britain and its colonies.[15] Locating excitement and anxiety about energy in literary, political, educational, and scientific texts demonstrates the cultural depth of electric power in the British world. Energy physics and popular interpretations of the theories that developed therein "had an undeniable appeal for those who sought one rule to rule them all."[16] This was especially true of natural philosophers and scientists like Herbert Spencer and William Thomson (later Lord Kelvin). For many, an all-guiding principle of the conservation of energy represented God, and for others promoted British liberalism and colonialism. That is not to say that every colonial politician or public intellectual wielded energy physics as a justification for the British Empire; rather, it resonated with British colonialism's ideas of uniformity, universal rules of law, and the domination of Nature.[17] Electrification gained adherents because it aligned so well with British colonial aspirations for New Zealand and empowered political authorities to argue for increased central governance, while granting Māori new ways to fight legislation and regain land.

## THE ROLE OF TECHNOLOGY IN THE EXECUTION OF EMPIRE

The British Parliament initially established the Colonial Office to manage British colonial affairs in North America. However, in 1854, Parliament restructured the office; now, all colonies but India remained under its direction, headed by the secretary of state for the colonies. New Zealand's colonization was not an isolated act of colonialism; it represented part of a global shift in Anglo population and culture, often termed the *settler revolution*.[18] In New Zealand, the settlement was continuous and complex. Settler and Māori efforts to use electric power systems to employ or alter colonial power make this struggle especially clear.

Throughout this book, I use the terms *colonialism, imperialism, settler-colonialism,* and *empire* as analytical categories to explore the history of New Zealand and its transition to electrification. These words are not synonymous. *Colonialism* is a practice of domination involving subjugation or exploitation of one person or group by another. *Imperialism* refers to the political apparatus that operates at the center of an empire. Colonialism can be an imperial act that necessitates invasion, whereas imperialism does not require land acquisition.[19] *Settler-colonialism,* distinct from colonialism, involves the displacement of Indigenous populations by settler society.[20] Settler-colonialism does not begin with the invasion of others' land, nor does it end with the dissolution of the imperial title. It is embedded in the legal philosophies and political institutions of countries worldwide. We must understand that while British settler-colonialism is separate from subject-colonialism (where settlers do not displace Indigenous peoples), the practices did coexist in New Zealand. The transition to electric power in New Zealand was both a colonial and imperial practice, creating a sociotechnical order that privileged white British settler society and their goals.

Colonial historians use technology as an analytic because the materiality of technology helps give substance to the sociopolitical and cultural abstractions and ideologies in play in colonial spaces.[21] Technology is of major import in records of the manifestation of imperial power because the British saw infrastructures, such as electric power systems, as a mechanism to spread material goods, knowledge, and culture into their colonies.[22] For instance, technologies offer concrete representations of British ideas about modernity, such as electric mining equipment or hydroelectric dams, because the British explicitly defined these technologies as modern. Electrical technology represents one of the foremost examples of British (and European) superiority for colonizers and some subsequent historians. European colonizers wielded technology for centuries as a principal justification for their movement onto and theft of others' lands. As Mary Pratt argues in her book about the imperialism embedded in travel writing, technology consistently reinforced European justifications for the empire.[23] Historians of technology have been complicit in continuing such motives by oversimplifying; in their narratives, they cite technology as a primary reason for successful European conquest and assimilation rather than dissecting how people used technology.[24] Historians like David Arnold and Arthur Asseraf have challenged these arguments by exposing the complex use and redirection of technologies intended to colonize.[25]

At the same time, focusing on technology can highlight flaws in settler narratives, as James Belich shows in his work on the New Zealand Wars.[26] Observing how colonizers and the colonized used technology lets us explore successful and failed colonial projects, thus challenging narratives of European social and technical superiority, and charting the clash between the imagination and the realization of colonial initiatives. Colonizers attempted to use technology to facilitate their visions and establish their superiority. Conversely, sociotechnological systems often change or perform unexpectedly and alter the visions of colonizers or work to resist colonization. Technology did work in some cases to exercise power, but more often than not, technologies and their users created new identities and versions of modernity to counter the intentions of European colonizers. Providing such nuance in imperial histories will continue to uncover the spaces of resistance to empire, allow for the histories of the colonized to be told in more detail, and challenge the ghost of determinism that still haunts historical studies of European colonial successes.

Although this book explores a subject that superficially appears to be a thoroughgoing colonial success—New Zealand did electrify rapidly—it highlights the unintended trajectories of these efforts over the long term. Electrification was far from an easy or obvious solution to New Zealand's desire for power; it was at times more difficult than planned to implement systems, or efforts failed to improve on existing systems. Although electrification was intended to enforce colonial authority, Māori often exposed or reconfigured colonial power dynamics built into New Zealand's electric grid.

New Zealand's electrification makes an excellent case study for historians of technology in the British Empire and those studying the relationship between technology and colonial power. Electric power was not an afterthought of British settlers who had already colonized New Zealand. Rather, politicians and engineers integrated electrification into colonial processes, such as communication, transportation, and resource extraction, bent on establishing British settlers' dominance over the Indigenous population and launching a self-sustaining colony.

The founding ideology of New Zealand's first settlers and politicians advantaged strong public oversight into the infrastructure of the first towns and ports. Although not all of these policies were implemented, they remain a constant refrain in New Zealand public works and industry. Driven by Malthusian fears of overpopulation and pauperism in Great Britain, the political theorist Edward Gibbon Wakefield (1796–1862), one of the many architects of a British colony in New Zealand, proposed a process of government intervention in the distribution of land in the colonies known as "systematic colonization," which he borrowed from Charles Buller's (1806–1848) speech of the same name to Parliament on the status of the British Empire.[27] Wakefield's system suggested that the right kind of emigrant might make the most use of the colony in service to empire and civilization itself. Others, like Robert Pemberton, similarly imagined and proposed that New Zealand offered bounties of resources and opportunities and promised a "happy colony" (figure I.1). He envisioned a "labour kingdom" in which the workmen of the world might be educated in the mechanical arts, philosophy, and the morality of hard work in towns both agrarian and modern.[28] Nowhere in their early nineteenth-century plans for the settler-colonies did Wakefield or Pemberton argue for implementing electrical technologies. However, the systems and interests they lauded laid the framing for the systems that electrified New Zealand. They stressed the significance of implementing modern technologies, extracting natural resources as efficiently as possible, and increasing centralized bureaucratic authority to create a British colonial state that would be immune to the industrial practices flourishing in England that led to political chaos

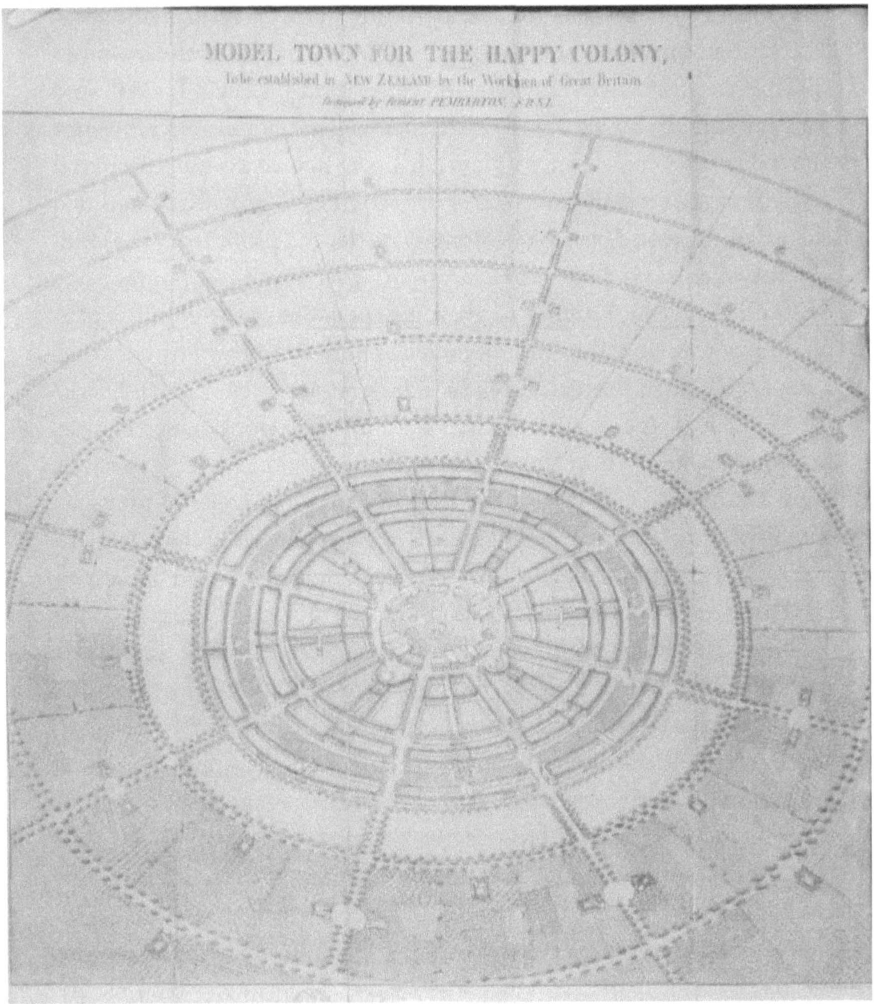


FIGURE I.1. Robert Pemberton, *The Happy Colony* (Saunders and Otely, 1854), appendix.

and depravity.[29] From the beginning, advocates for New Zealand's electrification replicated Wakefield's and others' foundational vision for a systematic and controlled colonial state. Without ever actually achieving it, the colony and state pursued systematic colonization.

The interest in electric power in New Zealand, especially on the part of settlers and settler government, was born out of other energy debates in Great Britain, especially in the advocacy for non-fossil-fueled electric power production in the late nineteenth century. These early debates demonstrate that electrical scientists and engineers did not consider fossil fuels the sole option for powering electrification. The British electrical engineering and booster communities viewed specific methods and structures of generation (i.e., wind, tidal, and waterpower) as fitting for British sociocultural values of efficiency and self-sufficiency—not to mention how they could quell imperial anxieties about energy resources. For settlers, self-sufficiency signified their hardiness and fitness to colonize, settle, and profit from the land. For the colonial (and later the Dominion) government, self-sufficiency denoted the colony's ability to grow independent of the empire's protection, administration, and financing. These conversations occurred across the British Empire; Canada, South Africa, New Zealand, British Honduras, India, China, and Japan all undertook major electrification projects within the formal and informal empire.[30] However, New Zealand's place within the British Empire as a site of promise, "the England of the Pacific," makes it an especially clear that electrification was not an inevitable next step for industrialization but a concerted effort by colonizers and the colonized to achieve political goals.

New Zealand's late-nineteenth-century electrification resembles Britain's, although the management and distribution of power differs, and advocates for electrification wielded the country's identity as a British settler-colony as a justification for adopting electric power technologies. For many, electric power and its requisite technological systems represented progress in Britain's long imperial project. Not every New Zealander identified electricity with the empire, but the industries they worked in adopted electric power and the cities they settled used public electric lighting. Electric power provided the means to conquer nature, efficiently distribute resources, and justify the creation of a modern (British) state—in other words, to meet the promises of British colonialism. Settlers used electrical infrastructures to facilitate their frontier settlement, expand their extractive industries, and legitimize a national government.

In New Zealand and elsewhere, most electrical history presents electrical technologies and the requisite knowledge to electrify as Western. Such perspectives on electrification privilege the success, dominance, and necessity of colonizers. In its intent and form, electrification was inseparable from the politics of the British colonization of New Zealand. The state wielded it through public works to extract resources, legitimize the Dominion, and gain energy security, and the Māori used it to navigate the colonial legacies embedded in

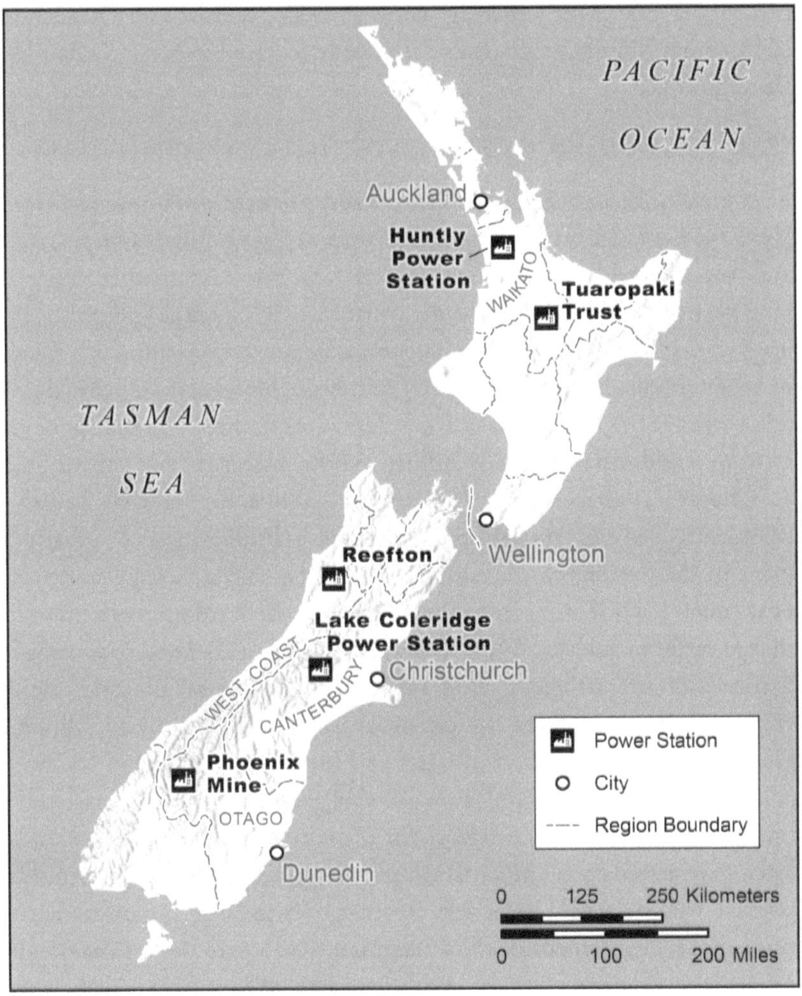


FIGURE I.2. Map of Aotearoa New Zealand with present-day borders and the locations of the power stations analyzed in this book.

the grid. Electric power infrastructures served colonial authorities and legislators by providing mechanisms through which the government could displace Māori from their lands. However, many Māori also used the same infrastructures for their own economic vitality, to reclaim stolen land, and to reshape the New Zealand's environmental policy. Much like Māori participation in war and the construction of New Zealander culture, as recent historians have emphasized, Māori laborers and urban users of electric power contributed to an intrinsic belonging to the nation.[31] Even though Māori narratives in Reefton,

Bullendale, and Lake Coleridge convey tales of removal, absence, and conformity, Māori designers and users participated in New Zealand's electrification.

## CASE STUDIES IN EMPOWERING COLONIALISM

This book investigates five electrification projects in Aotearoa New Zealand between the 1880s and 2000s, showing how each shaped and was shaped by colonialism (figure I.2). The case studies expose how deeply electrification became coupled with cultural and political initiatives in New Zealand and suggest that similar relationships exist in other spaces. By charting settler and Māori interactions with electric power for the last 150 years, I demonstrate how colonial policy, attitudes, and infrastructures influenced the process of electrification.

Chapter 1, covering the 1860s to the 1880s, argues that British colonizers used electric lighting in Reefton to fulfill visions of a settlement in the periphery, the superiority of British technology, and the possibility of a self-sufficient colonial town. British colonizers believed that in order to succeed, New Zealand settlements needed to be prosperous and self-sufficient while reflecting Britishness. Electric power promised to actualize the potential that politicians like William Gladstone, Julius Vogel, and Charles Hursthouse projected for the colony, and Reefton's population growth, gold mining prospects, and investors provided the framework for realizing the ideal New Zealand town. Animated by politicians, electrical engineers and salesmen, and settlers, the proposal to electrify Reefton enlisted existing infrastructures—namely, control of the Inangahua River—to create the ideal settlement. Engineers appealed to the town's interest in self-sufficiency, or its ability to generate a profit, by offering a technological novelty that provided a "never-ending" supply of power to light the town and power mining equipment. The electrification of Reefton demonstrates how New Zealand's earliest electric systems served colonial as much as local interests in lighting and power. Appealing to miners', settlers', and politicians' ideas about settlement, it became a model for settler-colonial initiatives in New Zealand.

The ability to use electric power in the frontier demonstrated the utility of lighting and fulfilled many imperial and settler visions for New Zealand. Electrification needed to fully align with British and settler plans for building a prosperous and self-sustaining colony, which included resource extraction and manufacture. Gold mines were the first industry to explore the potential of applied electric power beyond lighting. Advocates for electrification hoped to exemplify the aims of the model colony by increasing mining profits, conquering and using natural resources, and highlighting British technological superiority on an industrial scale. Chapter 2 follows one such project at the Phoenix Gold Mine between the 1880s and 1890s. In 1886, the mine implemented a hydroelectric scheme to power mining equipment. The installation demonstrates how electrification, particularly hydroelectrification, allowed for the redistribution of productive industries that served imperial and settler interests while conquering New Zealand's unique landscapes. Once a barrier to expansion, New Zealand's vast river system came to be used to justify the movement of New Zealand's industries outside urban spaces, allowing for industrial-scale production to align with British ideas of progress and New Zealander pastoral identity. Hydroelectric power at the Phoenix Mine transformed New Zealand from a Pacific resource frontier—a materially exploitable region in the colonial periphery—into a model colony ready for urban energy infrastructure.

Hydroelectricity proved an effective way to achieve monetary gain and establish colonial infrastructures that were used for lighting frontier outposts, mining gold in the frontier, and long-distance telegraphic communication. Chapter 3 charts the connections of electric power infrastructures to settler-colonialism in the establishment of New Zealand's first state-run hydroelectric works, Lake Coleridge Power Station, between the 1890s and 1920. Across the empire, the British enacted water regulations to control settler populations, marginalize Indigenous peoples, and facilitate new infrastructures. The completion of the power station at Lake Coleridge encapsulates the successful conclusion of the Dominion experiment and the origin of New Zealand's national electric utility, which continued to operate under the colonial models of resource manipulation, centralized political authority, and theft of Indigenous land and water rights in the name of technological progress and modernity.

Chapter 4, covering the 1950s through the 1980s, traces the expansion of New Zealand hydroelectric dependence and growth of the grid during the early twentieth century, leading to the decision to construct the Huntly Power Station. The Public Works Department wanted to build the station to expand the power provided on the North Island and create a more energy-independent state. The department's efforts to construct the plant in Waikato on Māori land circumvented new environmental regulations in order to meet the perceived demand for electric power. The Waikato Tainui resistance by way of appeals to the government to cease the project highlights the centrality of settler-colonialism in New Zealand because the power station was to be built on Tainui land. The state's priorities did not just work with hydropower but also extended to thermal power (steam from burning coal or gas). During the late twentieth century, the settler systems that shaped earlier schemes continued to guide the construction of New Zealand's largest thermal power station to date.

Last, chapter 5 investigates the impact and meaning of the Tūaropaki Trust's geothermal power station and the connected Māori-run industries. Following the success of Lake Coleridge, New Zealand's hydroelectric dams remained the preferred method of electric power production. In reaction to this pattern of electrification, particularly after the expansion of gas-fired stations, Māori Trusts used their legislated authority to reshape the country's energy policy by pursuing sustainable and globally just energy practices. Māori commitment to geothermal energy highlights the colonial structures still in place in New Zealand's electric grid, as well as the efforts of the Waitangi Tribunal.

In chapter 5, I argue that the Māori, particularly the Tūaropaki Trust, used and continue to expand geothermal electric power production to reclaim agency, establish a diverse Māori economy, and develop a sustainable energy future for New Zealand. Amid mounting concerns over climate change, geothermal energy has proven an effective aid to other forms of alternative energy by supplementing wind and solar farms during off hours. Unlike the electrical enterprises at Reefton, the Phoenix Mine, Lake Coleridge, or Huntly, geothermal energy in Aotearoa is rooted in Māori culture and energy practices. Māori investment, persistence, and environmental awareness in reclaiming Aotearoa's resources, such as those at the Mōkai Geothermal Fields, are the primary reasons for the success of these systems. The Tūaropaki Power Company's investments in geothermal energy, sustainable farming, and social justice represent an anticolonial effort to undo the exploitation wrought by earlier electric power systems in New Zealand. Through the Tūaropaki Power Company, Māori demonstrated how they could reclaim their lands, promote environmental sustainability, and reconceptualize colonial models of electric power production that deplete resources, are concerned primarily with profit, and emphasize centralization. The company's development points to how electric power companies can adopt ethical energy systems that can change, rather than replicate, colonial technological infrastructures.

## NEW ZEALAND HISTORIOGRAPHY AND THE ARCHIVES

This book aims to destabilize progress narratives that describe the islands' history as a quest for "national identity." Many of the islands' general histories, while including some Māori narratives, treat the history of New Zealand as the development of national identity that aligns with Western ideas about nationhood or citizenship. The quest for nationhood presupposes the teleological development from Polynesian homeland to a colonial outpost to an independent nation-state. Rigid understandings of New Zealand's national identity, particularly in the late nineteenth and twentieth centuries, stand as a euphemism for colonial systems of exploitation.[32] Confining New Zealand's history to the construction of a settler-nation also reinforces nationalist ideas that minimize outlying or problematic narratives. As has been well established by most postcolonial scholars, purely national histories focused on a single identity constrain the writing of history and reify the existence of the modern state.

Some authors in both popular and academic circles remain concerned with finding "the nation" or some unifying national identity emerging from colonial history.[33] However, since the early 2000s, historians of New Zealand have increasingly argued that more attention needs to be directed toward the islands' diversity of perspectives. Although violence, oppression, and theft feature heavily in New Zealand's history, they do not define it. The New Zealand story is not solely a settler story.[34] Instead, its history should be defined by the dynamic and mutually constitutive relationship between Māori and settlers.[35] We must understand New Zealand's history as a complex intersection of British culture, international settler movements, and Māori history. My work follows the example of historians like James Belich, Vincent O'Malley, and Linda Tuhiwai Smith—scholars who have moved away from assuming British primacy in New Zealand. Others, such as Felicity Barnes, Tony Ballantyne, Danny Kennan, and Manying Ip, have recently shown the history of New Zealand to be a dynamic and messy process in which many contested visions existed. The electrification of New Zealand helps highlight the contours of those dynamics.[36]

Historians of other British colonies, particularly Canada, Australia, and South Africa, have increasingly established historical frameworks that allow for transnational and cultural approaches.[37] A history of New Zealand's electrification depends on this historiography because electrical technologies and knowledge circulated throughout the British world including colonial and present-day New Zealand. Yet we must acknowledge the actors that resisted New Zealand's inclusion in the British world. In other words, we must shift away from making the British past the center of New Zealand's story. Most importantly, we must always engage with the lasting effects of colonialism.[38]

In this book, I work to situate New Zealand's electrification as part of colonialism rather than treating colonial politics as a precursor. Ballantyne, in *Webs of Empire*, says, "Colonisation and its legacies continue to stand at the heart of New Zealand life. The fact of colonisation is undeniable. But it is an awkward and frequently divisive heritage, generating fault lines across the cultural and political landscape."[39] Beginning with the fact of colonial rule, bureaucracy, and infrastructure is a necessary framing for understanding the history and impact of electric power systems in New Zealand.

Reckoning with colonialism in New Zealand's past requires engagement with a broad range of scholars, disciplines, and methodologies. In addition to incorporating and adding to the histories of energy transitions and technology and colonialism, this study builds on the methods and experiences of a broad range of Aotearoa New Zealand historians, including Māori, settlers, and scholars working from outside New Zealand (like myself). To best understand the technological systems I explored, it was most necessary to analyze the extent to which they intersected with displacement, violence, and erasure of the colonized in New Zealand, the Māori. Here, I have relied on the work of scholars like Atholl Anderson, Judith Binney, and Aroha Harris in *Tangata Whenua: A History*, Richard Hill's chronicle of the Māori, which relies on a wide variety of archival, archaeological, personal, and ethnographic research and the volumes on Māori–Crown relations.[40] Furthermore, I benefited immensely from the published and unpublished work of scholars at the Māori studies programs at the University of Otago, the University of Auckland, and the University of Waikato. New Zealand's historical journals, such as *The New Zealand Journal of History*, have increasingly included critical histories of colonization and the Māori.[41]

Postcolonial scholarship has certainly been useful for critiquing the colonial powers in action in New Zealand, but Kaupapa Māori scholars add that we must approach Māori history using Māori knowledge.[42] Kaupapa Māori refers to an ontological base that asserts an understanding of New Zealand's history as a partnership between the settlers and Māori. Tukuna Mate Nepe argues, "Māori society has its own distinctive knowledge base. This knowledge base has its origins in the metaphysical realm and emanates as a Kaupapa Māori 'body of knowledge' accumulated by experiences through history, of the Māori people. This Kaupapa Māori knowledge is the systematic organization of beliefs, experiences, understandings, and interpretations of the interactions of Māori people upon Māori people, and Māori people upon their world."[43] In particular, this scholarship demands that Māori perspectives, culture, and knowledge be applied to the history of New Zealand. This runs counter to many histories of New Zealand because it depends on critical theories developed by non-Western scholars. As an outsider, both non-Māori and non–New Zealander, and as a historian trained in a Western discipline, I aim to carefully critique and navigate the colonization of New Zealand through settler and Māori perspectives.

To complete the research underlying this book, I relied most heavily on Archives New Zealand and Alexander Turnbull National Library records. These archives are not simply repositories but an artifact of colonialism themselves. Since the beginning of British colonization, settlers generated written records to document, justify, and organize their cause. The colonial purview embedded in these archives and the records they contain must be observed whether poring over power station plans in Wellington or reading through another colonial governor's 1866 interpretation of the Treaty of Waitangi (1840). By viewing these archives within the larger formation of the British Empire, we can more clearly see the ways that networks and infrastructures, such as electric power systems or public works bureaucracies, shaped the colony and state of New Zealand.

Since starting this research in 2015, I navigated these archives physically and digitally to explore sources connected with New Zealand's electrification. Although I benefited from some institutional funding and circumstances that allowed me to search through New Zealand's physical archives, travel to power stations, and meet with people, I conducted significant portions of my research digitally, for reasons of efficiency and necessity. Both of the physical archives just mentioned possess excellent digitization infrastructures that helped me prepare for travel and supplement those findings with digital requisitions. Archival visits became impossible during the COVID-19 pandemic, which forced me to reexamine and repurpose much of my existing material and collect additional digital sources. This reliance on digital archives undoubtedly affected the range of sources I used. For example, I had expected to be able to facilitate more conversations with Māori scholars and enrich my methodological approach. At the same time, my reliance on digitization and the key archives honed my appreciation for the problematic power structures at play in historical research, such as travel, access, and publishing timelines. I have attempted to write that awareness into this book and hope that it aids future research in New Zealand.

Confronting colonialism in New Zealand's electrification narrative must also include an honest reflection on the archives that hold that history and how I encountered those sources. As many authors, such as Maria Fuentes, Ann Stohler, and Linda Tuhiwai Smith, have shown, colonial powers construct archives to facilitate the execution of their will. After all, New Zealand's colonial roots benefited from the systemic recordkeeping of census data and land records that privileged settler authority and minimized the presence of Māori in the archives. Such a power dynamic leads to archival gaps, omissions, and silences in colonial and noncolonial archives. These realities should not lead to historical silence. When confronted with a lack of sources, I worked to include Māori perspectives by describing colonial violence from the sources produced by settlers. The archival absence of Māori in relation to electric power systems helps prove the durability of colonial power embedded in that system. As a historian trained in archival research, I relied heavily on those sources. Although I did not employ the ethnographic work that might expand the Māori perspectives on electrification beyond the textual archive, I encountered extraordinary scholars, scholarship, and sources such as oral history repositories at the Alexander Turnbull Libraries. This is not meant as an excuse but rather as an encouraging statement about the work that remains to be done.

For this book, I included Māori perspectives on electrification and paid explicit attention to the ways that Māori were impacted (politically, physically, or socially) in the sociotechnical project of electrification. While Māori history has flourished in academia, accessing primary source material is often difficult or impossible. Almost all archival sources from the nineteenth century are settler in origin or come from the Native Lands Courts, which is not to suggest their illegitimacy or lack of utility but rather to acknowledge their limits. Moreover, many extant Māori records are not readily available to scholars. Therefore, it is often impossible to accurately depict—at least according to traditional methodologies—how Māori were involved in New Zealand's early electrification. Given these limits, I have included Māori wherever material exists and accordingly use as many Māori primary sources as possible. However, what is not recorded is in some ways as useful as the existing record. The chapters in this book in which settler voices outweigh Māori voices are not intended to silence or obscure the Māori. Instead, the absence of their perspectives should be an affirmation of the colonial intentions behind electric power infrastructures.

CHAPTER 1

# REEFTON

## NEW ZEALAND'S ELECTRICAL UTOPIA

The first electric utility in Aotearoa New Zealand was the telegraph. In 1861, George Grey (1812–1898), during his second governorship of New Zealand, ordered the installation of a telegraph line between Auckland and present-day Hamilton to communicate along the Waikato River.[1] The telegraph and the accompanying networks (including roads, railroads, and power lines) effectively introduced physical infrastructure that not only reinforced their colonial regime but also implicitly enforced ideas of land ownership and control that did not exist prior to invasion.

In 1879, the first recorded public display of electric lighting appeared, in a Wellington jewelry store, illuminating "Messrs. Kohn & Co" from behind a royal coat of arms.[2] The capital city of Wellington also saw the first demonstrations of practical applications of electric lighting. In 1881, journalists excitedly reported rugby and cricket matches lit by electric floodlights and speculated on electricity's potential. Two years later, Woodyear's Electric Circus lit up their center ring using electric power. According to onlookers, it was "brilliantly illuminated by electric light . . . as light as the day." Electric lights appeared on ships, such as the SS *Manapouri*, lit by 170 Swan incandescent lights. Members of Parliament visited the ship in June 1882, which prompted an electric lighting promotion in the New Plymouth House of Representatives building. Many of the components bore familiar names such as Siemens, Swan, Edison, and Brush. As one witty observer from the *Auckland Star* put it in 1881, "Electricity in Franklin's time was a wonder; now we make light of it."[3]

The electric light entranced colonial officials and settlers alike, who marveled at the sheer wonder and possibilities of electric modernity, including communication and power.[4] The public's excitement centered on visions of an electrically powered future to match the urban centers of Europe and the United States. Engineers and "electrically literate" entrepreneur-politicians, such as Alexander Bickerton (1842–1929) and Charles Lemon (1834–1901), published articles and advised administrators and businessmen of the potential of electricity.[5] New Zealand featured electricity in its industrial exhibitions and commercial advertising, much as the press and similar expositions had in Great Britain. Throughout the 1870s and 1880s, cities like Wellington and Christchurch held fairs that proclaimed electrical wonders. New Zealand newspapers printed many announcements for electrical exhibitions around the world from the 1860s onward. Given their wide circulation and success, New Zealand's papers kept citizens well apprised of electric power developments and excitement.[6] Boosters styled New Zealand exhibitions similarly to British ones, referencing the latter's technological might in the process. By the early 1880s, many settler towns, urban businesses, and industries were actively pursuing ways to implement electric power.

Around the same time, discussion about the potential of electric lighting and electricity as a motive force for industry frequently appeared in the press in Great Britain and its colonies. New Zealand was no exception. In the pages of *The Electrical World* and *The Telegraphic Journal and Electrical Review*, engineers and journalists mentioned new opportunities for electric power technology around the world. A single page of the February 25, 1887, *Telegraphic Journal and Electric Review* noted the *Great Eastern* steamship becoming a floating hotel; elevated electric trains in Elberfeld, Germany; the Sprague Company's development of a forty-two-foot electric bus; and the "Electric Transmission of Power and Lighting in New Zealand." The authors of that short piece claimed that "the antipodes are fully awake to the fact of electricity being an important industrial adjunct."[7] For British electrical engineers and many in New Zealand, successful electrification did not simply mean the presence of electric power technology; it meant the utilization of electricity to service the island and the British Empire's productivity.

In essence, global trends of the British Empire played out locally in New Zealand. During the 1880s and 1890s, New Zealand's newspapers resembled, if they did not copy, England's press. The papers published articles and advertisements touting the benefits of electricity and comparing the merits of electricity and natural gas works. For instance, the "Electricity Is Life" column from the *Thames Star* frequently published notes on useful electrical technologies like toothbrushes, hairbrushes, and a curious "flesh brush."[8] New Zealand did not merely replicate the excitement of Britain; its speculations and hopes for electricity connected to the ways it could benefit the colony, particularly protecting its status in empire through potential applications like lighthouses and military port illumination.

During these decades, New Zealand and London newspapers frequently paid tribute to the Reefton Electric Light Company. The hydropowered electrification of the town distinguished Reefton as a place of note in New Zealand and even within the more expansive British Empire. An anonymous author in the *Weekly Press*, a popular weekly newspaper published in Christchurch, commented, "Reefton is making a proud name for herself. She above all cities in the Southern Hemisphere, has started in the lead of utilizing the electric light for the benefit of her inhabitants. Magnificent Melbourne, stately Sydney, august Adelaide, with ambition Auckland, windy Wellington, critical Christchurch, and discursive Dunedin, are all left far behind in the march of progress, for Reefton has the electric light."[9] This remark is typical of the first mentions of New Zealand's electrification that I saw when scouring British periodicals from the 1880s for mentions of electric power technologies, such as tidal turbines and wind turbines.[10]

Reefton, or Reeftown, sometimes called Quartzopolis, is a small town on New Zealand's West Coast in the Inangahua River Valley, an area many settlers still considered the frontier during the late nineteenth century (figure 1.1). Settlers built Reefton during the 1870s to provide a service center for the Murray Creek alluvial quartz reefs. The great expanse of quartz (and its potential for gold deposits) required well-funded investors to purchase crushing and mining equip-

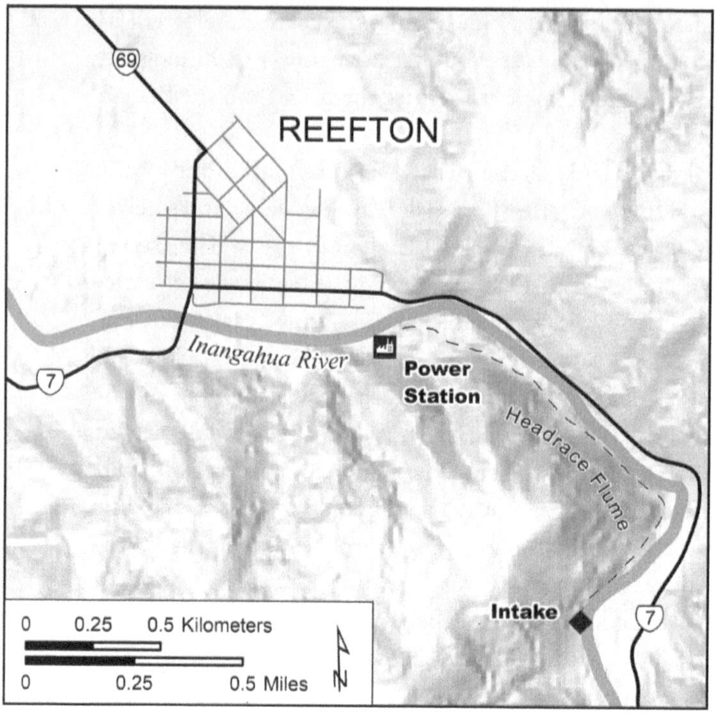


FIGURE 1.1. Map of Reefton hydroelectric scheme.

ment for the settlers to work. The difficult terrain, intense weather, and limited road networks made living harsh and transporting goods cumbersome. Still, the town managed to craft a cosmopolitan environment with settlers and laborers from across the world.[11]

Between the 1850s and 1870s, global gold rushes in California, Australia, and New Zealand sparked a wave of Anglo immigration worldwide, many thousands ending up in New Zealand.[12] Most settlers came from Eastern Australia and Britain. In addition to the so-called Anglo Expansion, other settlers from around the world appeared in large numbers in on the islands. Recently, historians have discovered evidence of a large Chinese immigrant population that has proved difficult to track because they avoided the official census (or were excluded by officials).[13] The settler population increased by 100,000 over the 1860s–1890s, thanks to the gold rushes and incentivized migration, enforced by politician-businessmen like Julius Vogel (1835–1899), known for his work as colonial treasurer and as premier of New Zealand. During this population explosion, mining towns like Reefton experienced massive growth and became more permanent settlements.

More importantly, the 1860s saw the creation of a settler society—a transition from the focus on short-term resource extraction to lasting relocation and population growth.[14] In New Zealand, "settler capitalists"—arrivals with capital to invest in businesses—and working-class migrants from England, Scotland, Australia, Germany, and China made up the bulk of the settler population. These settler capitalists were not of gentlemanly origins, although they saw themselves as such, and many became quite wealthy. By 1872, settlers, miners, and investors had constructed homes, hotels, stores, and banks, making Reefton a permanent home for some 3,570 citizens (others estimate it close to 4,000).[15]

In some ways, Reefton and the surrounding mining towns resembled more metropolitan centers in New Zealand because of the businesses and services.[16] The quartz fields of the Inangahua River, which ran through Reefton, measured forty miles long, and Reefton attracted hundreds of gold miners, merchants, and wealthy investors seeking profit by providing mining operations services. For example, Reefton maintained its own stock exchange for the quartz and gold prospects. According to local newspapers and regional histories, beginning in the 1870s, the entire community engaged in speculating.[17] In 1882, one reporter counted some 7,500 telegrams from the Reefton Stock Exchange. This period marks the highest level of activity on the West Coast and the sixth highest in the country.[18] As with most nineteenth-century boomtowns, the exchange made and lost fortunes. In 1886, some twenty-two citizens registered as share brokers or mining speculators. Popular stocks included those of the Welcome, Keep-It-Dark, and Globe mines. The Bank of New Zealand and the National Bank purchased the gold from and held the cash of Reeftonites. The *Inangahua Times* and *Inangahua Herald* supplied the town with daily news and the latest mining profit figures.

The high risk and speculative environment of Reefton fostered invention and attracted inventors and new mining technologies, including the Pelton wheel and hydroelectric systems. Reefton kept pace with the day's engineering and technological debates, such as hosting a lecture about the telephone:

> Mr. M. R. Gissinge of Reefton has, by the application of a very simple illustration in natural philosophy, been enable to convey a very fair idea of the telephone. . . . The "apparatus" consists of two empty wooden pill

> boxes and a piece of ordinary sewing cotton, say about twenty yards in length. A small opening being made in the bottom of boxes with a pin, the ends of the cotton are passed through each and knotted on the inside . . . it will be found that a person whispering in the box at the other end can be heard distinctly, the sound passing along the line precisely as a current of electricity may be supposed to do in ordinary telegraphy.[19]

Though remote and separated from urban centers and ports like Wellington, Christchurch, or Dunedin, Reefton's investors and business community saw in electricity the chance to assert a modern identity. Like many cities and towns in late nineteenth-century New Zealand and around the British Empire, Reefton drew the attention of settlers, entrepreneurs, and miners because of gold and its accompanying business potential.[20] The interest of Reefton's town leaders in electric power mirrored British enthusiasm about the potential of applied electricity in metropolitan spaces in greater New Zealand, Europe, and North America. The thriving mines played a major role in fueling Reefton's drive for electric power. In a town filled with speculative investors and businesspeople seeking profit from mines and miners, electric lighting and power systems greatly appealed to networks of New Zealand engineers, entrepreneurs, politicians, and journalists. These boosters or stakeholders of electric lighting and power systems constructed rhetoric that appealed to the town. For them, electricity promised to bring Reefton into the modern world.

The introduction of electric power added to Reefton's success and acclaim and is representative of how settlers imagined electric power would shape their ideal colony. Reefton's energetic residents thrived on the vibrant gold economy. Its predominantly single-male mining population established a need for temporary or transitional accommodation.[21] In the 1880s, the town boasted no fewer than seventeen hotels, where miners lived, socialized, and shared in regional brews, as seen in the photograph of Broad Street (figure 1.2).[22] With such amenities, Reefton appealed to engineers hoping to profit from electrical ventures. Accustomed to the risky ventures and new technology of mining, Reeftonites formed civic bodies to handle infrastructure proposals, and the town maintained enough capital to construct an electric system. However, neither the civic infrastructure nor the settler movements alone produced Reefton's electrification.

Like many other mining towns, Reefton's electrification differed

FIGURE 1.2. Reefton, by Henry Lock, as depicted in the 1870s. Purchased 1999 with New Zealand Lottery Grants Board funds. Te Papa (O.021408).

from the electric works of major cities like Wellington and Dunedin. Gold and other types of mining brought prosperity (or at least the idea that it could be achieved) into boomtowns throughout New Zealand. During the 1860s and 1870s, Reefton's gold industry had surged. To put the boom into perspective, in 1902 the *New Zealand Graphic* claimed that gold profits in Reefton exceeded £20 million.[23] But the boom slowed in the 1880s due to the necessity of deeper mining in the quartz reefs. The prospect of electrical infrastructures reinvigorated that sense of prosperity and progress in settlement, and Reefton continued to bustle with miners, businesses, and financial opportunities.

Amid this anticipation, in 1883, Walter Prince, an electrical engineer and entrepreneur, visited Reefton as he passed through West Coast settlements selling the idea of electric power. There he lectured on the benefits electricity might bring to Reefton. At his address to the Provisional Committee of Reefton, he claimed that "Reefton was singularly adapted for the introduction of a never-failing supply of electric power from its nearby river, especially compared to the existing gas system. It could be utilized for driving public machinery and powering regional mining equipment."[24] According to the local papers, the talk resonated so well that plans to implement such a system began immediately.

Prior to his first appearance in Reefton, Prince served as a consultant for R. E. Fletcher and Company, a Dunedin-based engineering and construction firm. Some historians suggest he worked for Julius Vogel's Electric Light and Power Company, but evidence of his prior employment is suspect until he appears in publications about Reefton's electrification.[25] Other than working for these companies, little record exists of his early work or his education. Based on the scheme he designed at Skipper's Creek for Phoenix Mine (see chapter 2), some historians have questioned his exact engineering knowledge and expertise. Still, Prince must have demonstrated some advertising and engineering skill, because he persuaded experts and nonexperts of his projects' validity.[26] According to one journalist, "Prince is essentially an electrician, or more properly, an electrical engineer, who, besides being an enthusiast in this new and wonderful science, has qualified himself practically by going through every grade of the profession in the celebrated electric manufactory generally known as the 'Brush Factory,' finally graduating as one of their most competent engineers."[27] Prince pitched the beneficence of electricity by offering a series of lectures and demonstrations. In 1886, while recovering from a horse-riding injury between projects in other cities, Prince stopped again in Reefton. He brought with him a one-kilowatt dynamo and persuaded local businesses to display electric lighting. Several hotels in town as well as Oxley's Brewery, a popular spot for miners, ran the lights. Over the next few weeks, Prince lectured and, with the strong support of the town, helped found the Reefton Electrical Transmission of Power and Lighting Company, Ltd. The company appointed Prince the head electrical engineer and contractor, of course, and sold stock through the local papers.

Prince's project for Reefton relied on hydropower, a design that fit the existing water power infrastructure in place there and at many of the surrounding towns. The style of mining and quartz crushing already necessitated a complex water flume system. The company proposed to divert water from the Inangahua River near a small settlement outside of town known as Blacks Point and carry it via a water race, two kilometers down to a point across the river from Reefton. Prince and his engineers estimated that this water volume could generate enough electric power to supply approximately five hundred

FIGURE 1.3. Reefton, 1904, Dunedin, by Muir & Moodie. Te Papa (O.036798). The powerhouse and flume are in the bottom left hand side of the photograph.

lamps, for street, business, hotel, and house lighting. The proposed power station would operate a twenty-kilowatt Crompton bipolar dynamo.

In developing and proposing his plan, Prince capitalized on a culture that placed great value in high efficiency and low cost. He estimated that the cost of a turbine for a hydroelectric plant would total £1,800, whereas a steam plant with the same hundred-horsepower output would cost £2,800. Furthermore, he argued that the "work expenses, cost of coal, engineer's attendance, and annual depreciation" for steam would be much higher, £750 per year, whereas fuel, transport of power, and attendance with the hydro plant would be minimal, totaling only £300 per year.[28] Other than the benefits of public lighting, Prince maintained that electric power systems were cheaper and more efficient because the mines could harness both the electricity and potentially the water power for mining. In regions like Reefton, this created more profitable and self-sustaining mining operations, which could dredge and crush more material for longer hours.

Some criticized electricity and pointed out the problems with Prince's system. Despite the water source's availability, hydroelectricity was by no means a "natural" or easy prospect for Reefton. One author in the *Timaru Herald* lamented that "there is something in this particular proposal which is not far beyond experimental."[29] The author argued that Prince was too ambitious and did not know enough about electrical science to design the system, and suggested that the costs would be much higher even if electricity eventually worked. While most of the complaints related to expense and the early inconsistencies with power supply, this was the case for other successful electric power ventures throughout the country, such as the mines around Skipper's Creek, Auckland's electric lights, and electric transportation systems (figure 1.3).[30]

The construction proved more difficult than expected. The company buried cables, rather than hanging them, because of a conflict with the telegraph company. Burying lines was a relatively new practice.[31] Workers needed to build two additional water tunnels to bypass the main bluff in town, and frequently had to repair poorly insulated underground cables.[32] In the face of difficulties with design, supply, and finance, the town and the rest of New Zealand remained committed to the introduction of electric power. The *Ashburton Times* reported, "At present, great as has been the progress made in the economical production of electricity, there is still admittedly a large amount of waste in the process of changing this energy from one form to another. But as we have said, the difficulties are being gradually overcome."[33] By the end of 1887, newspapers celebrated the progress on the flume.[34] Yet the actual opening in 1888 stalled, not uncommon for many nineteenth-century electrical ventures. First, the power lines had to be laid and connected, which was not completed until April 1888. The lamps for the street lights were not received until May. Despite Prince's initial cost projections of £1,800, by this time the expenses neared £6,000. By July, all of the components had arrived in Reefton, but the water race, which had remained stagnant since January because a massive tree broke the flume wall in Black's Point, needed repairs. Finally, on August 1, 1888, the power station began setting up for an initial test.

Despite the complications, Reefton's and New Zealand's press heralded the first electrification project as a real success. As he had before, Prince strongly encouraged a public exhibition, and the town happily obliged.[35] An arc lamp was placed on a tall pole near the station, and sixteen incandescent lamps lit the building's interior. Businesses held several public trials during the first week, and local and regional papers celebrated the venture's success. "The light was much admired and perfectly steady. The trial is considered eminently successful,"[36] reported one. Another said, "The success of the trial was complete in every respect, and it is needless to say Mr. Prince and the spectators present were greatly delighted."[37] The largest exhibition took place in Oddfellows Hall, on the main street of Reefton, where paying customers could see fifty incandescent lamps from across the river at Black Point. Ever the salesman, Prince played to their desire to be modern. He even took time to advertise Thomson-Houston arc lamps because he was a representative for the British company in New Zealand.[38]

Immediately following the trial week, subscribers began to be hooked up to the system. The company set a charge of £3 per light, per year. By September some 130 lamps were connected to the system.[39] For comparison, large early stations like Thomas Edison's Pearl Street Station in New York powered closer to 1,000 lamps, but 130 lamps fit similar small projects such as the station at Rijeka Harbor in Croatia or Mill Creek 1 in California.[40] Forsyth and Masters, a local hardware store, made the first permanent connection, followed by houses on Buller Road. Newspapers reported that people were charmed with the electric light; according to the *Inangahua Times*, "the era of kerosene was over"—meaning that people were reconsidering how energy ought to be generated.[41]

As often happened with the power stations of the 1880s, construction errors proved extremely problematic for the company. Prince came under pressure from the local press because the cables he had buried short-circuited due to poor insulation and inexpert installation.[42] At first Prince deflected the reports by criticizing delayed and faulty equipment. Before he was able to follow through on this plan, the *Inangahua Times* reported that he was "closing his business affairs in Reefton" and selling his 2,000 shares in the company.[43] Prince did leave, but the reasons given for his departure vary. Some newspapers suggest he went to Auckland to develop an electric tram car system.[44] Other sources say he moved to Thames, another gold-mining town with an active stock exchange and anxious investors. After that, Prince disappeared from the record, except for his death date and debt note in Dunedin.[45]

However inexpert Prince's effort and despite his sudden, somewhat suspicious departure, enthusiasm for electricity was by no means dimmed, which suggests that more than Prince's showmanship accounted for local uptake. By October, Prince had been replaced by John Joshua Horton, an electrical engineer, who continued connecting parts of the town and repairing the system. The *Inangahua Times* reported, "Most of the business places in Broadway [one of the main streets] are now lighting by electricity and, what with crowds of visitors, elegantly decorated windows, and the blaze of electric light, Reefton during the coming holiday evenings will make a considerable show."[46] Horton served as the Reefton Electric Company's head engineer for fifteen years, but because of constant repairs, especially to the water race, the company lost money.

Despite the occasional complaint, many in the town continued to adopt electric power and remained excited about its prospects. James Stevenson and John Dawson both connected their hotels to the electrical system. One of the hallmarks of "Dawson's Hotel" was being the first electrically lit hotel in the Southern Hemisphere.[47] Even Saint Stephen's Anglican Church, one of Reefton's many churches, connected to the lighting network. William Hindmarsh, who happened to be the electric company's secretary, was a church vestry member. The church received a custom rate, according to its minute book: ten lights on Sunday for thirty pence a week, three lights for any weeknight service for ten pence a week, and free lights any time the choir practiced.[48] The supply of electric power for lighting was becoming embedded in the town's basic functions.

Reefton's electric system delighted much of its citizenry and became an object of surprise and envy elsewhere, as is evident in a grumbling telegram from Sydney featured in the *Kumara Times*: "If there is one place on the face of the earth which is left all the time it is the city of Sydney. . . . At Reefton, on the West Coast of New Zealand—a poor little mining township—the electric light is laid on the streets, and private houses are supplied . . . while this forgotten wart is lit up as before with the sickly yellow glare of an expensive gas system."[49] Later in November, notice of Reefton's lighting trial appeared in the British publication *The Electrician* in an article titled "Progress in New Zealand," which proclaimed the settlement's success in lighting the whole town.[50]

## IMPERIAL IMAGINARIES

More than just population growth spurred the interest of Reefton's town leaders in electrification. As a comment from the *Grey River Argus*, one of the day's major West Coast papers, makes clear, it was also the imaginary they came to share about the place of Reefton and electricity in a modernizing New Zealand: "Electricity is coming to the front more and more every day. It is such an economical and versatile servant that it is being utilized in a variety of ways not dreamt of a few years ago. . . . It is making its way rapidly in the colony as an illuminant and a motive power; and electrical engineers of eminence at Home have their eyes on these Islands as presenting a fine field for the utilization of electricity in all of its protean forms."[51] The imagination of what New Zealand's future could be was opening places like Reefton to the possibilities of modernity, including electrification.

The architects of this imaginary, operating with many different agendas, maintained a belief in New Zealand's promise as a special place for the expansion of British ideals and culture. New Zealand offered resources and opportunities unavailable in other colonies (and Britain itself). Politicians and authors alike positioned New Zealand as the "Britain of the South," arguing that it could emulate the best parts of Great Britain, such as political institutions and world-leading technological progress, and remove the worst, such as overpopulation and economic corruption. Critiques of Great Britain's industrialization were loudest from the working class, Chartists, and early UK Liberal Party writers.[52] These self-styled visionaries spoke from London to New Zealand and reached an audience of equal breadth. By painting New Zealand as an "emigration field," with the potential to create an improved Britishness—or a cultural identity of Great Britain as home or center—and to become a self-sufficient yet inseparable part of the British Empire, this imagined ideal of New Zealand provided a framework that informed electrification efforts.

Historical case studies of gender, class, literature, and technology demonstrate the importance of understanding the conceptual spaces of the British Empire as an imaginary, used to define what is British, and, conversely, what is not. For decades, sociologists, historians, and literary theorists have used the imaginary to question and deepen understandings of the relationship between Britain and its empire. For example, David Graham and Jason Rudy show how advertisers and authors created spaces where settlers and officials could attach their hopes and aspirations.[53] Such scholars have consistently challenged the assumption that a unified vision characterized the empire; indeed, such an idea did not exist. Tracing the imaginaries at work in empire, they have studied how analysts can render layered pictures of modernity, imperialism, and British identity.[54] This explanation is not complexity for its own sake but rather a necessity, for the British Empire existed as much in military and legislative action as in the rhetoric of morality and progress used to justify it.

Many historians argue that one of the most formative periods for British identity happened during the nineteenth century because of the spread of British peoples and culture around the world. To transplant their culture and justify colonization, colonial authorities, settlers, and the Colonial Office collectively formed definitions of Britishness.[55] Addressing colonialism through the imagination grants agency to an expanded range of actors; the metropole and colonies produced mutually constitutive knowledge and experience, making older models isolating the settlement from the center increasingly suspect. Acknowledging the problems inherent in adopting terms such as *modernity*, I draw attention to how the British and New Zealand's settlers used concepts like modernity alongside technology to promote the British Empire. Michael Adas, Clive Dewey, and Leo Coleman have demonstrated that the British saw technology as a tool for achieving and managing a vast empire, whether that superiority existed or not.[56] The realization of that goal and the superiority of British technology rightly have been criticized. Nonetheless, understanding British imperial motivation requires charting their usage of technology as the key to modernity.

If we are to use imaginaries to tie together the empire and electrification, they must be collective. The imperial imaginary that concerned New Zealand was made as much by the British enactment of imperial power as by those settlers who portrayed New Zealand as the ultimate example of a thriving colony. Reefton's electrification performs a sociotechnical imaginary, a vision for the future that embedded electrification and empire within each other. I have borrowed the concept of the *sociotechnical imaginary* from Sheila Jasanoff, who defines it as "collectively held, institutionally stabilized, and publicly performed visions of desirable futures, animated by shared understandings of forms of social life and social order attainable through, and supportive of, advances in science and technology."[57] By framing Reefton's electrification this way, we add "social thickness and complexity" to the project by showing how a few individuals' shared ideas led to the communal adoption of a vision for electric power and what ought to be the dominant energy source.[58]

The Reefton Electric Light Company's plan for electrification fit into the broader imperial imaginaries formulated by politicians such as New Zealand's Premier Julius Vogel and British Prime Minister William Gladstone (1809–1898), each of whom anticipated the development of New Zealand into a self-sufficient colony, fit to become a Dominion, a "Britain of the South."[59] On one hand, it is an imperial imaginary because Reefton's electrification aligned with British interests to reduce colonial spending and military intervention while spreading British culture in the settlement. On the other, the town's transition to electric power performed the settler-colonial imaginary of local politicians, engineers, and the town's citizenry by demonstrating the colony's technological prowess and settlers' fitness to master New Zealand's frontier.

## Colonization, Settlers, and Māori

Although it will not surprise scholars that colonial authorities used electrical technology to expand their influence, the deployment of electric power in Reefton also exposes the many layers of colonial exploitation within electric infrastructure. From the introduction of electrical technology, the government attempted to use the accompanying infrastructures to advance the settler-colonial projects of supporting the non-Indigenous population or boosting the economic worth of the colony. As users adopted electrical technologies like the telegraph and saw the potential of others like the electric light, electric power systems became part of the British colonial arsenal. In New Zealand, the Colonial Office used electric power, much like other colonial public works, to take land, direct the economy, control resources, and encourage settlement. To meaningfully explore the relationship between electrification and settler-colonialism in Reefton and throughout New Zealand, we require some background on the relationship between the Māori and settlers as the latter colonized the land.

Before the British invasion of Aotearoa, the Māori and Moriori people populated the islands. The Māori lived on the New Zealand archipelago islands, whereas the Moriori inhabited Rēkohu, what became known as the Chatham Islands. Both cultures descend from Polynesian peoples that emigrated from Taiwan, Melanesia, and the Society Islands. Most anthropologists and historians place the settlement of New Zealand circa 1280 CE. The descendants of these settlers were the Māori. The separate settlement of Rēkohu by the Moriori occurred around 1500 CE. Each region developed distinct dialects, agricultural practices, hunting rituals, and warfare tactics.[60] Although I refer mostly to the Māori throughout this book because British legislation regarding electrification and land confiscation generally referred only to them, the term *Māori* comprises a vast and complex culture, not one homogeneous people. The Māori did not identify themselves as a unified political entity before the European invasion, and many Māori did not recognize negotiations with the Europeans as universal. The category's continued usage as wholly representative has led to cultural, legal, and philosophical conflict.[61]

In 1642, Abel Tasman (1603–1659) landed near Golden Bay on the South Island's northwestern tip. Māori attacked his crew, killing four and leading Tasman to name the place "Murderers' Bay."[62] The remaining European sailors left to survey the other island. In 1645, the Dutch designated the islands the *Nova Zeelandia* after the Dutch province of Zeeland, which James Cook (1728–1779) anglicized in 1770 to New Zealand. Cook began surveying New Zealand in 1769, under the guise of studying Venus's transits, while he planned to make the islands a British territory.[63] He returned to New Zealand on all three of his significant voyages (1770, 1773, and 1778), improving the region's mapping and negotiating with Māori for goods and guidance around the islands. Not to be outmaneuvered by the British, French explorers also sought territory in New Zealand. In 1772, Marc-Joseph Marion du Fresne (1724–1772) charted Spirits Bay and the Bay of Islands. After many successful exchanges, further miscommunication and theft led to the murder of du Fresne and the massacre of his crew. While tragic for both the Māori and French killed during the conflict, this massacre's legacy resulted in more devastating ramifications for the Māori. Reports of the event circulated throughout Europe, giving the Māori a misleading reputation for being more dangerous for settlers than settlers were to Māori. Decades later, the memory of these encounters fueled violent resentment toward the Māori in legislation and warfare.[64]

In 1788, New Zealand and Australia officially became part of the colony of New South Wales. By the 1790s, New Zealand had become a trading hub for British, French, and American whalers and merchants. Settlers traded metal goods, such as tools and guns, for Māori food, water, flax, and sex. Most reported peaceful negotiations, but occasionally conflicts arose. From the beginning of British settlement, New Zealand's Anglo society assumed an air of superiority as a British colony that would surpass earlier colonies, most notably Australia. Historians refer to the early portrayal of a superior New Zealand as the "Wakefieldian Dream," after British colonial theorist Edward Gibbon Wakefield (1796–1862). During the 1820s, Wakefield advocated for "systematic colonization," which included buying land from the Māori, paying high wages to settlers, and incentivizing passage to the colony from Great Britain.[65] Over time, colonial authorities and settlers adopted many of Wakefield's views, especially his resentment toward non-British immigrants and the Māori. His plans for a structured and primarily middle-class settler population became the foundation for many visions of New Zealand's future.[66] The New Zealand Association (1836–1839), later the New Zealand Company (1839–1852), and the Crown government enforced the idea that New Zealand would remain devoid of the "stain" of association with the transportation of British convicts as was the case with Australia. New Zealand represented a new moral standard for the British Empire to attract an adventurous and financially capable middle class.[67]

However, all the talk about settlement assumed and relied on the removal or decline (real and imagined) of Māori populations. Settlers could move into these areas only if the Māori moved or were removed. For colonial officials, an ideal New Zealand necessitated the banishment of the Māori. To convince settlers and financiers of the colony's potential, they needed to create "empty spaces." The colonization and settlement of New Zealand depended on the appropriation of Māori persons, land, and resources. While historians traditionally have treated this as an uncomplicated triumph, the story is more complicated. As James Belich argued, while historical successes remain controversial, the certainty of a British victory has rarely been questioned in the historical record. During the nineteenth century and in subsequent historical works, the British treated the Māori removal as a single concentrated effort, when no such operation existed.[68] Despite European portrayals, the Māori were a culturally and politically diverse group and an integral part of New Zealand's colonial society. The reception and interpretation of the European invasion varied. Many iwi (the broadest subdivision of Māori, best translated as "nation") welcomed trade and relations with Europeans, while many ignored or resisted their presence on the islands.

In the decades before the 1830s, the number of British settlers increased from a few hundred to around a thousand, primarily engaged in trading and missionary work. By 1840, the Māori population had shrunk from 100,000 to between 50,000 and 80,000, whereas the non-Māori population had risen to over 2,000.[69] The high rate of Māori population decline in New Zealand furthered the settler-colonial idea of New Zealand's potential, in contrast to the "troubled" settlements of South Africa and Australia or the brutal colonization of India. The extent to which British settlement contributed to Māori population decline remains a controversial and unresolved topic.[70] Despite British proclamations of total success, many Māori rejected British authority or superiority and succeeded in rebuffing both the British military and their legal administration. Yet over time, Māori did lose considerable land and rights to aggressive settler and military campaigns.[71]

British politicians and settlers portrayed the decreasing Māori population as a consequence of the Māori's barbarous traditions and justified their colonization on this basis. For many of these settlers and politicians, New Zealand's resistant Indigenous population, geographic separation, and status as the last-touched land on Earth, *terra nullius*, justified the beneficent civilizing mission of the British Empire, a mission run by its sober middle-class settlers. During the 1830s and 1840s, British authorities and settlers rhetorically located New Zealand at the physical and racial edge of the world. Ernst Dieffenbach (1811–1855), a German naturalist with the New Zealand Company, argued that the Māori "are a people decidedly closer in relation to us, than any other; they are endowed with uncommonly good intellectual faculties; they are an agricultural nation, with fixed domicile, and have reached the farthest point of civilization which they possibly could, without the aid of other nations, and without the example of history."[72] Still, settler-colonialists feared that the absence of British guidance might lead to horrific war and a decline in Māori progress toward civilization. New Zealand Governor George Grey (1812–1898) worried that granting Māori leaders too much power would "draw back the mass of the native population to their barbarous custom."[73] This isolation of the island and misleading stories of Māori ferocity fabricated a terrifying "other" to juxtapose with British civilization. Only by removing, conquering, and assimilating the Māori and Moriori, believed colonial boosters, could they demonstrate the colony's success to potential middle-class British settlers, other British colonies, investors, and competing European nations.

In 1840, the Colonial Office and settler representatives signed a Royal Charter and separated New Zealand from New South Wales under the direction of William Hobson (1792–1842), New Zealand's British consul. Shortly after that, Hobson coauthored the Treaty of Waitangi (1840), which established British sovereignty over New Zealand under his governorship and recognized Māori land ownership, giving Queen Victoria's government the sole right to purchase that land.[74] British representatives drafted the treaty, and five hundred Māori leaders signed. Today the Treaty of Waitangi is considered the founding document of the state of New Zealand. However, in London and New Zealand at the time, British legislators and Māori interpreted the treaty's language differently. Settlers understood the treaty as a declaration of total British sovereignty, which granted the governor the right to rule all of the island's inhabitants. By contrast, the Māori broadly believed the treaty ensured a sharing of sovereignty with the crown in exchange for military and legal protection.

The Treaty of Waitangi remains one of the most controversial documents in New Zealand history. Questions remain about whether it led to the New Zealand Wars (1845–1872), a series of armed conflicts between the Māori and New Zealand government. What is clear is that the British government and settlers abused land purchases and participated in violent movements against the Māori after the treaty was signed; that is, settlers superficially adhered to the treaty's terms because the motivation behind colonization was not recognizing Māori rights but taking their land.

After the New Zealand Wars ended in 1872, the British implemented sweeping land confiscation to punish the Māori for rebellion. This practice breached the Treaty of Waitangi and shifted the settlement pattern in New Zealand because settlers could more easily acquire Māori land.[75] Postwar colonial officials and settlers increasingly regulated extractive and production industries, including logging, introducing new plants and animals, and exporting meat.[76] For instance, as part of Vogel's plan to bolster the colony's longevity, Parliament passed the Forest Act 1874 to establish state forests and prevent deforestation, not for conservation but for later profit. Settlers used the laws surrounding Māori land ownership and the low price of land to convince or force Māori to move. By purchasing land in bulk from Māori owners, settlers and the colonial government could sequester Māori iwi, removing Indigenous access to such communal resources as water.[77]

## Visions of New Zealand in the Empire

Although they never explicitly wrote about Reefton, William Gladstone, Julius Vogel, and Charles F. Hursthouse (1817–1876) all heralded the potential of New Zealand, each contributing something to the collective imaginary taken up by electricity advocates. Their perspectives represent an imperial and colonial imaginary constructed around New Zealand where British imperialism and colonial self-sufficiency thrived. (Within the British Empire, *self-sufficiency* referred to a colony's ability to self-govern and remain economically stable.) In the words of a contemporary settlers' guide, the colony reflected "everything of England, in short, but the soil."[78] Colonial officials and writers based their aspirations on the social and resource potential of New Zealand, the efficiency and spirit of settlers, the conquest and assimilation of the Māori, and belief in their colonial mission.[79] They based this sociotechnical imaginary on the political biases of New Zealand's colonial government, the self-sufficiency supplied by the country's resources, and British technological superiority. Electric power promised to be the ideal means to meet these expectations.

Gladstone, British prime minister and chancellor of the exchequer, saw the possibility to establish the ideal colony in New Zealand. In the more extensive empire, history remembers him for his commitment to "home rule," the idea that colonies ought to be ruled by their citizens, especially but not only Ireland.[80] His commentary on New Zealand's Second Constitution in 1852 summarizes his imperial plans for New Zealand: "My belief is, that the highly intelligent community you have founded in New Zealand more thoroughly reflects the spirit, the character, and intelligence of England than almost any other among all your colonies, and my opinion is, that they will exercise so clear an intelligence. In discerning what is for their own good, that they will rectify the error of our crude legislation, and will, extricating themselves from this complexity, attain to a unity of system and a clear and accurate discrimination of power."[81] Gladstone's ideas about New Zealand resembled those expressed by most British politicians. The Colonial Office, and many politicians in New Zealand, believed that the colony provided a model superior to the older subject and settler-colonies. Governor Grey exclaimed that "Her Majesty's Government may rest satisfied that there will be soon no more prosperous or contented Settlements than those which have been established in the Southern District of New Zealand."[82] New Zealand could avoid the stigma and troubles of earlier colonies, like Australia and South Africa, because of the types of settlers moving to the island and the government's displacement of the Indigenous population.

Gladstone's use of the word *reflects*—that New Zealanders reflected England's spirit, character, and intelligence—is noteworthy. By *reflects,* he meant constructing or duplicating Britishness. Authority figures in Britain and New Zealand, like Gladstone and Grey, maintained that the islands would become the "Britain of the South" or the "England of the Pacific." They imagined New Zealand as a place for the British, not the Māori or non-British immigrants. New Zealand settlers strived to maintain their unique Britishness through writing, advertising, and eventually film from the end of the nineteenth century and well into the twentieth.[83]

As a politician, author, and entrepreneur who served as premier of New Zealand from 1873 to 1875, Julius Vogel's speeches and published works offer the clearest examples of the sociotechnical imaginary politicians constructed around electric lighting. Before his premiership, Vogel played a significant role in the formation of the Australasian Electric Light, Power & Storage Company and its subsidiary, the New Zealand Electric Lighting Company, Ltd.[84] Furthermore, he advocated for New Zealand's federalization and a particular vision of its place in an expansive empire. In 1874, he wrote to James Ferguson, the governor of New Zealand: "I hope that the British possessions will in time be consolidated into a mighty Dominion, and the British Colonies become, not dependencies, but integral and inseparable parts of an Empire."[85] Note especially his use of the term *dependencies.* New Zealand's potential resided in the qualities that could make it a less dependent and more productive part of the empire. In 1878, still vying for a change in New Zealand's status, Vogel published "New Zealand and the South Sea Islands, and Their Relation to the Empire" in the *Proceedings of the Royal Colonial Institute*, in which he argued, "No land on Earth has before it fairer promise. The bond people share there and the resource potential provides [*sic*] a way to maintain common interests."[86] Vogel further detailed the people and islands' resources to argue for the colony's federation. The bond and common interest he spoke of is the "reflection" of Britishness that Gladstone described. The term *resources* refers to many exports, but also to the country's energy potential, such as its rivers, to which he referred repeatedly. Vogel described New Zealand as having the potential to be the ideal product of settler-colonialism, both in its ability to maintain British culture and politics and through its landscape's prospects for self-sufficiency.

Others besides politicians and boosters like Vogel advocated for increased settlement. Many settlers composed guides to life there to encourage people to move. One of the more famous of these works is Charles Hursthouse's *New Zealand: The Britain of the South* (1857). Notably, he sought to attract the right kind of British citizens to the "natural gifts" of New Zealand and prevent them from going to America or Canada. They could leave overcrowded England and find a spacious but civilized society in New Zealand.[87] Hursthouse called himself a New Zealand colonist and a former visitor to the United States, Canada, Cape Colony, and Australia, but he wrote and promoted this book while living in London.[88] The book's focus on the individual settler obscured his motivations—namely, that he was paid by the New Zealand Company. During the 1850s and 1860s, Hursthouse composed a lengthy argument for moving to New Zealand. Some historians find that his work merges history with "propaganda and prophecy."[89] While Hursthouse's work cannot be cited as an accurate portrayal of real opportunity in New Zealand, the sentiment and imaginary he constructed are undoubtedly representative of how the British government and travel writers promoted the colony. Hursthouse's work, and the many others like it, were widely read and likely played a significant role in influencing settlement to New Zealand and, more importantly, settler expectations for the colony. Settlers expected growing industries and productive landscapes, particularly those newcomers who made their way to mining settlements like Reefton.

To understand the imaginary that many in Reefton would take to heart, it is worth noting the chapter in Hursthouse's book where he explores New Zealand as a "mineral kingdom." He quotes earlier surveys of the country, speculating that New Zealand likely would provide vast mineral wealth, including gold, coal, iron, and copper. He mentions Governor Robert FitzRoy (1805–1865), most well known for his role as captain of the HMS *Beagle* on Charles Darwin's famous voyage, and his prediction that "beneath the productive surface of these teeming islands are mineral stores as yet hardly known."[90] Hursthouse stoked his readers' imaginations by arguing that New Zealand offered "the world's most precious ores and metals; and they have sometimes lain, almost under our very eyes, undetected for years. It has been so in America, in Africa, in Australia. . . . Hundreds of shepherds roamed the plains of Victoria [Australia] for a quarter-century little dreaming that they walked on gold. And though no country has less need of mineral wealth than New Zealand, a few years may prove her as rich *below* the surface, as she is above and on it."[91] Hursthouse's argument captures the hopes many settlers and investors held for New Zealand during the last half of the nineteenth century. It sidestepped arguments against the search for gold by citing other gold finds beneath colonies after decades of occupation. For Hursthouse, establishing New Zealand's mineral wealth meant stabilizing the economy and promoting a self-sufficient colony. Although his writing predated the introduction of electric power in New Zealand, the imaginary he constructed for New Zealand's future harmonized with policies that mining towns like Reefton adopted to promote their longevity as an economically viable (and maybe essential) part of the empire.

New Zealand's colonial-potential ideals established a discourse that echoed in the rhetoric around Reefton's electrification.[92] In the case of Reefton and New Zealand more broadly, the press and electric power boosters attached ideas of self-sufficiency—both economic and political—and a craving for technological prowess to colonial visions for New Zealand. A passage from an 1869 New Zealand Institute lecture demonstrates that for New Zealanders (and the empire), technology secured and justified empire: "Electricity and steam communication appear as the appointed agents of an All-wise Providence for building up the comity of nation;—for obliterating prejudices;—for throwing down restrictions on free intercourse in trade, science, literature, and all the generous amenities that should bind man to man."[93]

As its settler-colonies expanded in the late nineteenth century, New Zealand colonial officials saw the promise of an economically productive self-governing nation like theirs as justifying their model of colonialism. Electrification thrived because hydroelectric power not only lent itself to settler visions for New Zealand but also fit within broader settler frameworks such as using Māori land for projects seen as modern and establishing more industry outside of urban centers.

Compared with England's rural spaces, New Zealand's rugged topography, like that surrounding Reefton, was seen by settlers as needing the right kind of technological improvements and colonial vigor. Yet settlers and British-colonial boosters feared replicating the worst aspects of Britain's industrial developments, including urban pollution and slums. By the late nineteenth century, electric power, especially hydroelectric, promised to bring proper urbanization, meaning none of the dirty smokestacks or massive factories that resulted in objectionable working conditions.[94] Many of New Zealand's resource practices from the earliest days of settlement adopted a rhetoric of pastoral utility, meaning settlers hoped to harness resources efficiently without falling prey to destructive industrialization realities like poverty and overcrowding. New Zealand's settler government framed the colony's burgeoning industries, like mining and the production of dairy, wool, and meat, as a mechanism to critique overindustrialization and exhaustive spending in England—to become a "better Britain. Even though the actual development of these industries resulted in poverty, the exhaustion of resources, and unsustainable population booms, the pursuit of hydroelectric power became inseparable from the drive to build a better Britain because it enabled the affordable production of motive power for rural industries.

Prince's successful proposal emphasized the ways that electricity, especially hydroelectricity, could produce greater self-sufficiency. He stressed, first, the issue of a "never failing supply," a phrase that appears in most of Prince's presentations on his hydroelectric system.[95] The promise of a never-ending supply of energy echoes many discussions of the future of electricity in Great Britain, particularly in the arguments of physicists and engineers such as C. W. Siemens (1823–1883) and Lord Kelvin on the potential of batteries.[96] Some physicists and engineers believed batteries to be a promising large-scale means of generating electric power. The talk about unending supply was connected both to local hopes for self-sufficiency and broader concerns within the empire about the potential loss of energy sources, something that plays out in terms of a discourse about energy efficiency. For example, William Stanley Jevons (1835–1882), an economist and logician, and John Ruskin (1819–1900), a writer and social theorist, wrote about the limits of industrialization or imperial resources.[97] In 1882 the *New Zealand Times* offered, "In both [New Zealand and Australia] the stores of coal and forests of timber, comparatively speaking are scarcely touched. Let these new colonies take warning . . . once dug out of its bed, it can never be replaced—wood will grow, coal will not."[98]

Hydroelectric power, already gaining steam in Great Britain and its colonies by the 1880s, seemed a promising solution. During the 1880s, papers in New Zealand noted the successes of hydropower potential at places like Niagara Falls by citing the works of William Siemens or industrialist and engineer William Armstrong (1810–1900). Noting that the power potential would be extreme, one advocate wrote in the *Lyttleton Times,* a major Christchurch paper, that "all the coal raised in the world would be scarcely sufficient to supply" the power contained in the waters of Niagara Falls. The same author also noted that electric power produced this way would afford remote towns disconnected from other energy systems to industrialize. Towns like Reefton could develop power and lighting systems that would be done "in a superior manner to anything that has ever been done by gas."[99] Unlike more famous hydroelectric exhibitions like Niagara Falls, Reefton showed how other colonial settlements might generate electric power. In New Zealand, the prospect of an endless supply of energy in the form of hydropower moved the colony away from a dependency on coal or installing an increasingly expensive natural gas supply network and toward steam-driven power stations.

Prince's system also promised to help produce a colony that did not financially strain the empire, through economical production of electric power. Prince himself argued, "The great aim of mankind has always been, since the advent of steam, to find an economical means for the subdivision of power."[100] He would argue that this meant moving toward a subdivision of electric power production to several small-scale hydroelectric works instead of having production in one place. He therefore aligned with electrical specialists in Great Britain, like Sylvanus Thompson or James Blyth, who argued for decentralized arrangements using wind turbines or secondary batteries to electrify nonurban spaces.[101] While not explicitly connected to the electrification of Reefton, this British discourse strongly influenced broader imperial policy. The practical and moral anxieties about energy efficiency spoke to the role of sustainable energy supplies in expanding and maintaining the empire.[102]

Electric technologies allowed settlers to adapt British ideas about technological modernity while forging their image of self-sufficiency. Reefton's electrification provided lighting and attracted additional investors, but the project succeeded because it promised a form of economic self-sufficiency, productivity, and permanence. The town's acceptance of Prince's proposal and its resonance throughout the empire demonstrate how this project worked to perform the sociotechnical imaginary of figures like Vogel. For both Reefton and New Zealand, Prince tapped into the desire for self-sufficiency expressed in Vogel or Gladstone's imaginaries while simultaneously appealing to the investors and settler population. The decision of Prince, the investors, and the town to use hydropower stemmed as much from the desire of the colony's leaders and entrepreneurs to be self-sufficient as it did from the geographical availability of water. They made hydropower fit. Thus, New Zealand's first recorded use of hydroelectric power—Reefton's—developed not because of the convenience of the Inangahua River but because of settler ideals of self-sufficiency. Reefton's choice of water for generating electricity was not inevitable, convenient, or even economical—but New Zealand and the international electrical engineering community both took notice.

## A TOWN OF LIGHT

Reefton sits within the Paparoa and Victoria ranges next to the Inangahua River, which the settlers called the Thackeray River before 1866. Today, Reefton's mining and settler past is apparent if you walk down Broadway Street, which turns into New Zealand State Highway 7. You can stay at the Lantern Court Motel, its exterior restored to resemble a nineteenth-century hotel, or sit and listen to the actors at the Bearded Mining Company regale you with tall tales of their mining days and Reefton's history while you drink Billy Tea. And if you are lucky, you might find gold. In this way, Reefton typifies the mining heritage of dozens of towns around New Zealand and Australia. What distinguishes Reefton from the others that use mining to yoke themselves to the tourist industry is the phrase attached to

FIGURE 1.4. Emma Timmis's mural on a Chorus (internet provider) box in Reefton, March 15, 2023. Photo credit: Emma Timmis, used with permission.

the town sign: *Reefton, "The Town of Light"*—a slogan that now also adorns murals and even the town's website.[103] Even in 2023, people cherished and celebrated the imagined and real electric town. Emma Tillis, who completed a mural on a Reefton telecommunication box (figure 1.4), said, "As Reefton is the town of light, a bulb and beams of light spread through the design with golden rays to represent gold mining history."[104]

Of more interest on a stroll through Reefton is the absence of the deeper history of people who called and still call the Inangahua River home—the Waitaha, Ngāti Māmoe, and Ngāi Tahu, all Māori iwi.[105] Reefton did grow into one of the West Coast's most famous mining towns and today highlights that history, the electric light, and some of the best mountain biking on the South Island. But those things only exist because of the violence wrought upon the people there by the

British Empire, its settlers, and their desire for resources. The written record, produced by settlers, leaves little information for us on the Māori there at the time. Yet the silence and written obfuscation of the Māori exposes the settler project in Reefton. While the settler visions of electric power help us weave colonialism more clearly into the early foundations of New Zealand's electrification, not everyone saw the land and resources there this way. We may never be able to fully reconstruct how Māori miners felt or what Māori communities did in the years immediately after Reefton founded the power company. However, we do know that in 1956, in Reefton and the surrounding mining villages, the Land Development Committee published a report on the status of the land for living and agriculture—and were mortified. The land had been made barren and unusable by the mining and dredging operation in which Reefton flourished and electrified. The only arable lands in the area, according to the report, were Māori lands, such as Matai Flat.[106]

Reefton gained international fame for its implementation of a public, electric lighting utility. Taken out of the British Empire's context, Reefton's electrification appears only as a curious aside in the global transition to electric power, a token curiosity in the nineteenth century's electricity craze. However, framing Reefton within the larger context of empire connects the transition to electric power with broader colonial aims. Powerful imperial rhetoric, colonial policy, and technological idealism influenced Reefton's acceptance of electric power. People in Reefton made it happen, and they did so before the British Empire established electric power as a utility. In the press, and subsequently in the town's twentieth-century tourism efforts, the harnessing of electric power forged a distinct community amid so many other mining towns. The town's experience highlights how New Zealand's electric systems served both imperial and local interests in lighting and power—a point I build on the work of other scholars, such as On Barak and Dipesh Chakrabarty, who use both technology and the "periphery" of the empire to show that, however remote, Reefton's story of electrification extended beyond local generation and use of electric power.[107] It was part of a broader global interest in electric power and British settlers' desire to modernize the frontier.

Walter Prince's proposal to electrify Reefton using a "never-ending" supply of power to light the town and run mining equipment amalgamated with the visions of an ideal colonial settlement. Even though many of Prince's promises faltered, and he likely sought profit over the greater project of transitioning to electricity, the boosters behind Reefton's electrification believed electric power would benefit their town as a mining settlement and as a part of the larger British Empire. Reefton and New Zealand fit an imperial imaginary forged through decades of colonization and increased settlement. To succeed, New Zealand settlements needed to be prosperous, self-sufficient, and reflect Britishness, which in turn meant adopting modern technologies. Reefton's population growth, gold mining prospects, and investors created the correct climate to introduce electric power. Its story thus becomes a useful tool for appreciating the imbrication of colonial power into the physical infrastructure and conceptualization of New Zealand's—and indeed other British colonies'—electric grids.

## CHAPTER 2

# MINING FOR POWER

## ELECTRIFYING THE PHOENIX MINE AND THE CREATION OF A MODEL COLONY

Electricity in places like Reefton excited people, but lighting only partially demonstrated electric power's colonial utility. For the transition to electric power to succeed, boosters needed to implement systems aligned with British and settler plans for building a prosperous and self-sufficient colony. Introducing modern British technology to the frontier, as had happened in Reefton, was the first step. Next, investors and settlers sought to apply electric power systems to improve New Zealand's economic worth to the empire, both for settlers' profit and to cut Great Britain's cost of maintaining the distant island colony.

Among those at the forefront of exploring the potential of electric power application beyond lighting were mine owners. In 1886, the Phoenix Gold Mine implemented a hydroelectric scheme to power mining equipment. As the builder and user of one of New Zealand's first electric power systems, the Phoenix Mine represents how settlers came to rely on hydroelectric power, often at the expense of the integrity of waterways, to generate power in the service of the empire. Electrifying the mine's equipment promised to increase company profits, use natural resources, and highlight British technological superiority—aligning with British settler-colonialism's larger aims.

Though few found riches in New Zealand's tumultuous gold-mining frenzy, the rush jumpstarted its economy. Both settlers and Māori had participated in the California gold rush in the 1840s and brought many mining methods back to New Zealand.[1] Diggers searched for gold washed along riverbeds (alluvial), on beaches sparkling with flecks, and on coastal terraces that required sluicing and deep shaft mines for the gold locked in quartz reefs. Frequently, the natural manifestations of gold determined the type of town they established. Sluicing and dredging rivers meant more temporary or mobile establishments, while the deeper mines tended to lead to more permanent towns because of the equipment and skilled workforce required to maintain it. In 1862, following a rush in Otago, triggered by a rare find of eighty-seven pounds of gold from Dunstan Gorge, hopeful diggers spread out around the region near present-day Cromwell on the South Island. Many moved to the area around Skipper's Creek, a tributary of the Shotover River. Miners quickly found numerous alluvial quartz reefs, the largest of which was named the Scandinavian Reef.[2] The deep quartz deposits indicated the likelihood of deep gold veins, prompting a desire for a permanent mine early on, which set the stage for the Phoenix Mine.

In May 1864, miners for the Scandinavian Quartz Mining Company, named for their other work on the Scandinavian Reef, erected at Skipper's Creek a four-stamp crushing battery—a type of milling machine used to break down material by pounding rather than grinding—powered by a water wheel on the creek's right-hand branch.[3] The battery crushed ore into sand, which was then washed and screened to remove pyrite and unwanted sediment. Miners then mixed the sand with mercury to create an amalgam that could be separated into pure gold.[4] There were numerous types of stampers, many of which were invented by engineers in the United States during the California gold rushes. New Zealand miners adapted much of their mining technology from US and Australian mining equipment.[5] The battery at Skipper's Creek was of the "Cornish" style, developed in Cornwall in the 1850s for tin mining.

In 1866, after little success with surface mining, the Scandinavian Mining Company ordered new equipment from an Australian company, A. K. Smith's Carlton Foundry.[6] The new system consisted of thirty stamps powered by an overshot water wheel (figure 2.1). This new battery remained in service until 1907.[7] According to George

FIGURE 2.1. The Phoenix Mine battery.

Henry Frederick Ulrich (1830–1900), a geologist of the Royal Society of New Zealand, "it was the largest in the Province, and its system of gold-saving appliances resembles most closely that of the Port Phillips Company, Clunes. The yield of sand by this latter process has varied from 3 to 12 oz. of gold per ton."[8] However, water shortages, which resulted from drought and sharing the river with other mining companies, kept the mine from operating at its full potential. Furthermore, the fumes produced from the crushing were hazardous to the men working around the battery.

The mine struggled to stay open until it was purchased by George F. Bullen (1832–1912) and his brother in 1874 and renamed the Phoenix Quartz Mining Company.[9] The Bullen family established the mining settlement of Bullendale, sometimes called "the Reef" or "Skipper's" by miners or contemporary maps. During this period, newspapers consistently reported progress and regional excitement about the prospects for the Phoenix Mine, especially after Bullen's takeover. By 1884 the mine was producing 145 ounces per week.[10] In the intervening decade, waterpower reliability proved troublesome for the mine, which led to a search for alternative power sources. Spending nearly £15,000, Bullen expanded the mine and added several new instruments, such as a Leffel turbine, which harnessed falling water to power the battery.[11] Irregular waterfall nonetheless plagued the mine's productivity. Beginning in 1884, Bullen and his mining manager, Fred Evans, pursued the electrification of the battery.[12]

Bullen hired the Dunedin-based firm R. E. Fletcher and Co. to supply materials and install the electrical equipment. Walter Prince—the same Prince who advertised and helped design the system in Reefton—was appointed the mine's acting advisor.[13] Company records and mine inspections credit Prince as advising engineering, but there were likely others involved. Whatever his inadequacies were as an engineer, Prince succeeded as a salesman and certainly played a role in winning Robert E. Fletcher (1857–1935) the contract. Fletcher and Co. also served as the local agent for Julius Vogel's Australasian Electric Light, Power, and Storage Company of Sydney and London, which had been responsible for the electrification of Sydney and Melbourne.[14] Besides Bullendale and Reefton, Fletcher installed many of the electric works in New Zealand. For instance, in 1901, it established hydroelectric works at Rotorua, the Okere Falls Power Station. It was the first plant that received some funding from the state rather than private enterprise.[15] The company was a logical choice to install the system at the Phoenix Mine.

The Phoenix Company and Fletcher and Co. began constructing the powerhouse on a flat near the left-hand branch of Skipper's Creek (figure 2.2). The mine works had operated on the right side, but the left branch had a steadier flow, and electric power could be transmitted over a distance, which circumvented the problem of water shortages. A water race, called Aylmer's Race, ran above the flat for an older mine works, which builders repurposed for the hydroelectric scheme.[16] They constructed the powerhouse at the base of a ridge near Skipper's Creek that sat 182 feet below the water race and brought water into the house via penstocks (figure 2.3). The water turned two Pelton wheels, which Fletcher had acquired from Price and Sons of Thames. Invented by an American engineer to drive gold mining equipment in California, the Pelton wheels signal the global exchange of engineering knowledge during the late nineteenth century, enabled by colonial settlements and resource extraction opportunities such as gold rushes.[17] Once the powerhouse had been constructed, Fletcher

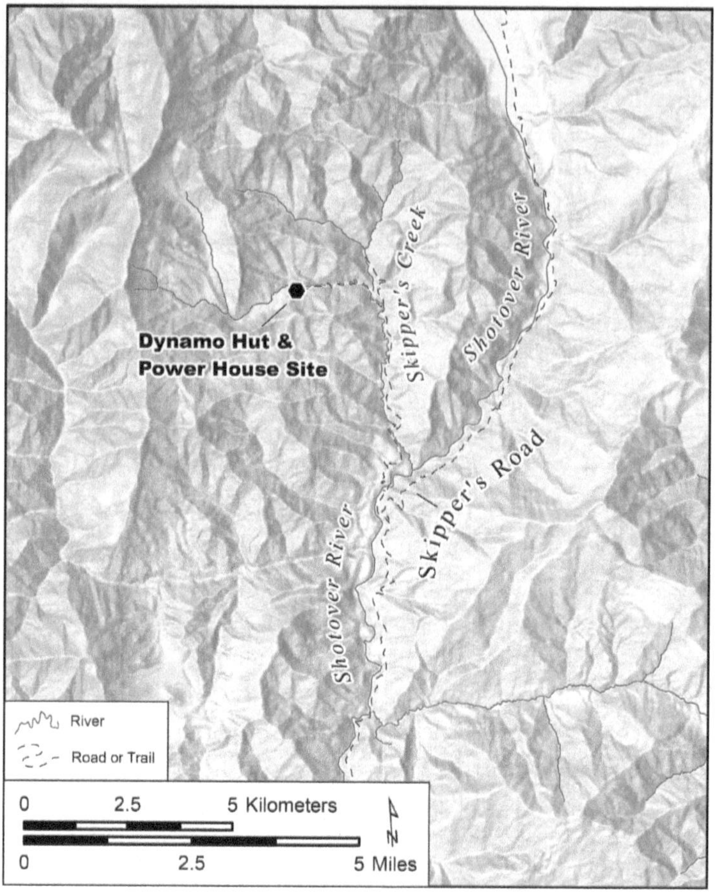


FIGURE 2.2. Phoenix Mine and powerhouse locations.

installed two forty-light arc dynamos, which turned the Pelton wheels and transmitted electric power to a six-pole Mordley-Victoria Motor located at the battery, affectionately named "Victoria." Engineers linked the dynamos and motor by No. 8 Birmingham-gauge copper wire, the 0.165-inch wire that was an industry standard at the time. Industrial archaeologists such as Peter Petchey have pointed out that the particular Victoria motor used at Bullendale was designed to be a dynamo, not an engine. During the late nineteenth century in New Zealand and elsewhere, electrical engineers often reconfigured motors to work as generators for lighting systems, even if the device was meant to power other equipment. For instance, engineers equipped the nearby Mosgiel Woolen Mill with a Victoria dynamo in 1885 for electric lighting. Such technological adaption became commonplace at mine works across New Zealand and highlights investors' and engineers' willingness to try anything.[18]

In February 1886, the system trialed successfully and drove twenty of the stampers.[19] At first, the mining community did not show much excitement about the installation because it merely kept pace with the power output provided by water, but this was only temporary.[20] Hopes remained high that electricity would improve the mine's productivity, so the Phoenix Company added a new rock breaker and more stampers. In 1887, Parliamentary Mines Inspector George J. Binns, a fellow of the Geological Society of New Zealand, made a highly detailed report of the mine's operation. Binns noted excitedly that the Brush dynamos were the "largest size yet manufactured," and mentioned that the powerhouse and battery had telephones so that the attendants could communicate.[21] By the end of 1888, the dynamos could power all thirty stamps.

The success of the project sparked interest around the country. Newspapers in New Zealand were littered with excitement and praise. Engineers and journalists frequently cited the success of the Phoenix Mine to encourage the installation of electric works in other spaces. "With regards to the prospects of the Phoenix Mine, we are happy to learn that they are exceedingly good—better than usual. The battery is turning out from 60 to 70 ounces of gold per day."[22] These pronouncements served two purposes. First, they encouraged investment in the mine by showcasing long term profits in the mine; and second, they connected those positive numbers to electric power. "New discoveries of payable stone continue to be made in the Phoenix mine [*sic*], enhancing the value of the mine considerably. The electric apparatus has been working with the greatest ease and regularity."[23] In 1886, citing the Phoenix mine, one miner complained of the mines near Reefton, saying, "Let us take the Keep-it-Dark, Mountain Maid, Main Lode, Lady Fayre, Premier, and New All Nations. Here six companies have literally no crushing power, saving by the sparse amount of water supplied during the summer. . . . A dynamo motor (like the one at Bullendale) could be driven by the Arrow River all year round . . . and keep their respective mines clear of debt."[24] Electric-powered mining equipment provided the promise the return on investment typical of just about every segment of the mining industry, but it also appealed to something so key to the settler-colonial designs for New Zealand, namely self-sufficiency.

FIGURE 2.3. The powerhouse at Skipper's Creek.

As this report makes clear, electric power distinguished this mine as more than just a successful business venture. Reported one newspaper, "The enterprise displayed by the owner Mr. G. F. Bullen of the Phoenix mine [*sic*], Queenstown, New Zealand, is an example deserving of imitation. . . . Of course, Mr. Bullen and his manager were looked upon as misguided enthusiasts . . . but now that the scheme is an accomplished fact, they will receive, no doubt, the congratulations they deserve."[25] Written six months after the mine's electrification, this remark captures the legacy of the Phoenix Mine in the historical record. It was a technological marvel that improved the mine's productivity and output. However, the phrase "misguided enthusiasts" points to the doubt and uncertainty surrounding the electrification process when it was first undertaken. Though many may have been quick to highlight the mine's use of electric power, others, especially those in positions to evaluate the mining industry on a grand scale in the colony, questioned the practicality of such an untested technology. For example, Binns, the parliamentary mines inspector, noted shortly after the electrification of the Phoenix Mine was completed, "Very little is yet really known about electricity," and admitted that he had no experience with electrical equipment.[26] The hydroelectric system did not necessarily make the construction any easier than other systems Bullen could have used. Workers, mainly using mules, moved tons of unassembled parts and wire two hundred miles to the mining site. Moreover, the mine engineers themselves lacked the expertise needed to install such a system successfully. For example, reports show that the cast iron component used in the Brush dynamos at mine had problems with eddy currents, loops of electric current created within conductors, resulting in loss of power between the dynamo and stamper that likely plagued the system's reliability.[27] Yet rather than continuing to use water to power mechanical stampers, the owners elected instead to construct an entirely new and unproven system.

Engineers and subsequent historians have noted the risk of investing in such an experimental and, therefore, potentially financially disastrous technology as electricity. Some scholars, such as Petchey, provide more nuanced histories of the mine's transition to electric power and place it in the context of the global gold rush and international collaboration of electrical scientists and engineers.[28] Other histories of New Zealand's electric power systems argue that the electrification of the Phoenix Mine was motivated primarily by the difficulty of transporting coal through the region, although in making this argument they oversimplify the building of a dam in Otago. However, the coal argument does not fully explain the choice. The mines could have continued to use or update their water-powered equipment, as most mines did. Electrified gold mines existed in Australia, South Africa, and the United States, especially for lighting, pumping water, amalgamation, and explosive ignition. Still, they did not power the sizable equipment like that at the Phoenix Mine.

Most did not begin to use electric power for such equipment until the mid-1890s.[29] Explaining the mine's transition to electric power, both in its successes and failures, is only possible by contextualizing the project as part of British colonial aspirations for New Zealand. Its integration of electric power should be recognized as part of British settler-colonialism: transforming colonial resource extraction into a self-sustaining industry that fit specifically with the ideal of New Zealand as a model colony, one that transplants a British population to the frontier by displacing Indigenous peoples, utilizes the land in service to colony and empire, and establishes the colony as financially self-sufficient.

## MAKING AN EFFICIENT COLONIAL SYSTEM

Hydroelectric power at the Phoenix Mine played a role in transforming New Zealand from a Pacific resource frontier—a materially exploitable region in the colonial periphery—into a model colony that did not financially burden the Colonial Office.[30] Even though the Phoenix Mine was isolated and far from the urban centers of New Zealand and indeed the British Empire, miners and owners remained connected to conversations about electricity. As Petchey shows, the Phoenix Mine's physical remnants help demonstrate the global circulation of electrical technologies and engineers, a valuable approach given the lack of records kept at nineteenth-century gold rush operations. However, focusing only on the technological marvel devolves into debates about primacy and sidelines the Phoenix Mine's legacy in New Zealand's existing, primarily hydroelectric, electrical infrastructures. It was the first mine in New Zealand to adopt electric power, but the implementation of electricity did not make it an outlier or ahead of its time. Rather, the decision to use electric power coalesced with British plans for a model colony, especially in the extraction of resources using modern British technology.

### Making Minerals More Malleable

The establishment of a model colony, one that exemplified how a settler-colony should develop and operate, prevailed as a consistent theme between the various imaginings for New Zealand.[31] Colonial authors used the phrase "model colony" in speeches, fiction, and travel guides. In an article using the phrase as its title, *The New Zealand Journal*, the mouthpiece of the New Zealand Company—which above all focused on providing a positive outlook on the colony for its local and British readers—made clear that New Zealand's success and perception of productivity were inseparable: "Production in New Zealand will, therefore, not be delayed as it was in South Australia; and the success of the experiment will be much earlier determined. The first colony at Port Nicholson [Wellington] will be a Model Colony; and will hereafter be referred to as such."[32]

From Cook's voyages forward, New Zealand's bountiful land and resource potential defined its value to the British Empire. Throughout the nineteenth century, social theorists like Edward Gibbon Wakefield (1796–1862), John Robert Godley (1814–1861), and Anthony Trollope (1815–1882) saw and wrote about New Zealand as a "resource or commodity frontier," one that would materially and socially enrich the British Empire, so long as settlers brought with them British culture, religion, and work ethic. Wakefield believed that settlers from the laboring classes, who could not usually afford emigration on their own, would make the best candidates because they could become landowners, establish an honorable "British" work ethic, and avoid the ruin caused by class division.[33] Wakefield argued that other colonial efforts had been fruitless and horrible because settlers with capital came and brought an attitude of "master and servant" toward those with less. To him, this divide resulted in stagnation wherein the population squandered the initial investments on pitiless labor, and no one took advantage of the opportunity to advance. Wakefield's "Systematic Colonization," and by extension the New Zealand Company's colonial plan, hinged upon the settlers' eventual ability to purchase land or own businesses that used the land, like mining companies or farms.

During the 1830s, the goal had been to establish New Zealand as another regional trading post for existing and future Pacific territories. However, by the 1840s, New Zealand offered an ideal place to redirect the ambitious British laboring class to settle, a new trading market, and a potentially resource-rich landscape. During Great Britain's Industrial Revolution (1760–1840) the introduction of new technologies, expansion of the British economy, and population increase created many problems in England. Many problems purportedly solved by industrialization became worse. Between 1800 and 1860, England's population grew from approximately 7 million to 18.3 million, stretching the country's available land and resources.[34] Increasingly, workers developed urban spaces like London and Manchester to take factory work, displacing farmers and sometimes skilled workers.[35] The demand for material resources, population control, and increased wealth stimulated British international initiatives to settle new territory, establish a global trading network, and compete with other European powers.

Between the 1850s and 1880s, colonial boosters published detailed, expository tracts pushing British citizens to move to New Zealand. The Colonial Office employed writers to praise the colony's merits. In these guides, authors pontificated over New Zealand's virtues. Its climate was better, it was a small island nation (and a potential naval authority like Britain), it would not have the "convict stain" of Australia, and the settlers would come from the middle class or higher. Arthur Clayden (1829–1899), author of *England of the Pacific* (1879), believed that the laboring classes, disenfranchised by the overcrowding of Great Britain, or facing job loss due to the growth of factories, would succeed in New Zealand by farming or extracting its mineral resources.[36] Both Clayden's work and Charles Hursthouse's (1817–1876) *New Zealand: The Britain of the South* (1857) show how the British thought about and advertised New Zealand's potential for vast mineral wealth as an incentive. For example, Hursthouse claimed that "from the peculiar formation of the country, coal, copper, and gold and the precious metals will probably be found in abundance."[37] Clayden, in *England of Pacific*, exclaimed:

> The abiding impression left on their minds was that the country was pre-eminently a grand and glorious one, one which not New Zealanders only, but every citizen of the British Empire might be justly proud of. It had clearly all the elements of prosperity about it, and more nearly resembled the promised land of the Israelites than even Canaan itself did. It was indisputably "a good land, a land of hills and valleys; a land of brooks and rivers; a land of sunshine and of song; a land whose stones are iron;" its sand along the sea-shore at Taranaki being literally of iron, and out of whose hills you may not only "dig brass" but gold, silver, copper, tin, marble, and "all precious minerals."[38]

Despite their inaccuracies and exaggeration, these books were widely read and went through numerous editions. Such texts promised prospective New Zealanders a country with untapped mineral potential, which appealed to those seeking to do the mining, build the towns,

FIGURE 2.4. New Zealand centennial stamp. New Zealand Post Ltd. Te Papa.

and invest in companies and relevant infrastructure. Collectively, they represented the imaginaries and visions constructed by and for New Zealand within the British Empire. Provided the correct tools and work ethic, they promised, New Zealand would yield mineral wealth and support a self-sustaining colony (figure 2.4).

## Rushing to New Zealand

Boosters and the Colonial Office continued to push settlement—so long as settlers came from the middle or laboring classes and worked to improve the profit of the colony—until the 1890s. More than any other resource or incentive from the Colonial Office, mass settlement during the 1860s and 1870s was encouraged by gold. Gold meant potential capital for settlers and the British Empire. In other colonies, especially Australia, gold had added to the imperial treasury. Compared with many so-called failures in the "subject colonies," the gold rush in Australia, Canada, and New Zealand generated profit for the British via higher taxation rates.[39] Building on global gold fever, the promise of mineral wealth bolstered New Zealand's growing reputation as a model colony. Moreso than wealth, gold brought an enthusiastic population of primarily British settlers to tame New Zealand's frontier. The Colonial Office readily paid to send settlers to establish permanent industries. If British settlers arrived excited to buy land or work in profitable industries like gold mines, permanent communities and civic infrastructures could flourish.

The New Zealand Company began advocating for increased mining during the 1840s and 1850s even though the prospect of mineral wealth remained speculative at the time. Continued investment was then legitimized by the scientific corroboration of gold in New Zealand.[40] Ferdinand Ritter von Hochstetter (1829–1884) conducted the most famous geological survey between 1858 and 1863 as part of the Austrian *Novara* Expedition.[41] During his wider expedition around the globe, Hochstetter spent time in New Zealand exploring volcanoes and the geological features of the island. In his book about the trip, *New Zealand: Its Physical Geography, Geology, and Natural History* (1867), Hochstetter said, "I was convinced, that if properly worked, the goldfields of the Aorere and Takaka Valleys near Golden Bay would prove very productive and that the discovery and working of those first goldfields of New Zealand would be followed by that of new goldfields extending all along the mountain-range of the South Island. The discovery of such fields, I was certain, would in the course of few years secure for New Zealand increasing importance amongst the gold-countries of the world."[42]

Once news of gold deposits circulated in Great Britain and the United States, Australia, and China, settlers poured into New Zealand in a series of gold rushes. Much like California, Australia, and South Africa, New Zealand witnessed a massive influx of settlers seeking to strike gold beginning in 1842 after whalers discovered a small quantity near Nelson on the North Island's Coromandel Peninsula. However, major mining operations did not start until 1852. The first significant gold rush to New Zealand, sometimes called the Otago Rush, began in 1861 when Gabriel Reed struck gold in that region. For the next couple of years, Otago became the center of interest in gold. Settlers, and sometimes Māori, panned for gold throughout Otago and founded numerous towns. Between 1864 and 1867, the gold prospectors moved to the West Coast of the South Island. Miners set up small camps during the first two years, but as mining continued more permanent settlements followed. Historians estimate that some 195,000 gold-seekers came to New Zealand during the 1860s, leaving a settler population of about 114,000 at the end of the decade, up from 2,000 in the 1840s.[43]

## Publicizing the Periphery

During the mid- to late nineteenth century, cost remained one of the most pressing logistical realities of colonization in New Zealand. Travel guides and reformist authors frequently omitted cost from their colonization plans. Colonial officials, however, did not. To attract settlers and justify the imperial investment in protection and trade, the colony needed to be profitable. Profitable resource ventures suggested the colonial government and settlers could establish a self-sufficient colony that could contribute to the high costs of maintaining a remote settlement and model colony.

Over the 1850s–1870s, the North Island mines produced £4.5 million, a much smaller yield than the gold mines on the South Island.[44] By the 1880s, the days of panning were over; most of the remaining gold operations required substantial investment in equipment and labor. Mine owners wanted to produce enough to attract investors and laborers, who in turn could dig deeper and process more quartz. To establish a long-term operation, miners needed sustainable methods of powering equipment. Many mines did not turn a profit because variability of local water flow could not keep their stamping batteries working. Concerns were expressed about wasted energy for dredging, sluicing, drilling, and crushing at mines worldwide. These waste concerns were not about efficiency but rather maximizing the use of resources present.

During the late 1880s, following Reefton's successful demonstration of its lighting scheme, the potential for hydroelectric-power generation built excitement. For many owners and investors, hydroelectrically generated electricity promised to avert the waste of energy and replace steam-generated electric power. Unlike hydropowered or steam-powered mining batteries, in New Zealand and elsewhere, hydroelectric systems took full advantage of the water running through mine works. At Reefton, on one side of the creek, water entered penstocks to turn the turbine. The other branch of the creek collected water to clean crushed material for faster processing. According to the designers, hydroelectric power more efficiently harnessed the energy from the waters around the mine. They argued that electric power could yield greater profit for the owners and miners, reaffirm the floundering gold rush, and establish a more permanent mine.

Some mines that opened during the gold rushes often stayed open for a year or less. In contrast, the Phoenix Mine remained in operation from 1862 to 1907.[45] Even though Binns's report did not show that electricity immediately raised profits, he hoped they would increase as the electric system improved. According to Fred Evans, at the end of 1886, the introduction of electric power increased the amount of material processed at the Phoenix Mine and attracted more investors. Evans argued:

> It may be well said that I have been well provided with funds for many years past in opening and prospecting the Phoenix Mine. I admit that, yet at the same time I can say that all of many drivings [*sic*] on lode or crosscut for so lengthened a period, there are none that are not of advantage now. But it cannot be said that I have copied anyone, either in mining or machinery, for I venture to think that in both we have left all others behind, and have shown the colony the immense value of its resources both in mining and power.[46]

In addition to attracting regional attention, the Phoenix Mine's electrification "attracted the attention of the Home Country."[47] International newspapers and journals took notice of electricity to mining. Of the many energy concerns at the end of the nineteenth century, electric power generation was of obsessive interest to engineers throughout the British Empire. During the 1880s, many questioned the longevity of steam as the motive power of the world, especially its reliance on coal.[48] The empire needed to power its navy and its production while maintaining its economic growth. During the European imperial entanglements with the Ottomans prior to World War I in the Middle East, oil offered an alternative.[49] For many, electric power, especially when generated by water, offered a superior power source. Thus, for owners and investors working in other industries, who sought to apply electric power to more than lighting, such as meatpacking or dairy farms, the experimental system sparked excitement.

The exact reasoning behind Bullen's pursuit of electric power for the mine remains unknown; he never documented his thinking. Some hints indicate Bullen's personal experience played a role; for instance, in 1885 he successfully had electric lights installed at the mine.[50] Nonetheless, the rhetoric around electricity speaks to an atmosphere of excitement about its promise that may have nudged investors like Bullen to use it in the hopes of increasing the productivity and value of New Zealand's frontier. Like so many other settlers, Bullen almost certainly would have read claims that electric power would increase profits by providing more efficient and locally sourced power. It is also likely that Bullen had heard of other successful electrification projects, specifically hydroelectric works around New Zealand such as Reefton's. By the time Bullen expressed interest in electric power systems, the company's owner, Robert Fletcher, and his partners in Dunedin represented part of a small hub of electrical engineers in the city, which connected the mine to a much larger apparatus of electric technopolitics. Electric mining technologies and professionals flourished in the city and supplied projects, like the Phoenix Mine, with expertise. From the 1860s and through the rest of the nineteenth century, Dunedin was home to many electrical engineering firms.[51] The New Zealand Electric Light Company operated there, too,[52] and in 1878 the University of Otago established the Otago School of Mines in Dunedin.[53]

In addition to linking the British Empire via electrical communication, sites of electrical knowledge production and invention connected many countries undergoing electrification. Electrical and civil engineers, especially from Europe and the United States, traveled around the world consulting on electrical projects, affording many a means to jump-start a career outside of the competitive European job market. Notable examples included British engineers like Richard Henry Brunton (1841–1901) and William Ayrton (1847–1908), who established electrical engineering colleges in Japan, and Anthony Walton White Evans, an American engineer (1817–1886), who developed railroads in Chile and New Zealand.[54] Prince and Fletcher had both been trained by the Anglo-American Brush Corporation based in London. A separate firm, the Anglo-American Brush Electric Light Company, had offices and engineers circulating between the United States, Canada, Mexico, Chile, England, New Zealand, and Japan. The company instituted this global engineering network by training its engineers and distributing them between projects where they sold Brush components.[55] Effectively, Brush's company helped physically tie colonial states around the world through extractive industry and communication networks. In the late nineteenth century, most considered Brush equipment, particularly dynamos, state of the art. Prince was one of the Brush engineers in New Zealand, where mines almost exclusively used Brush components.[56]

Advocates for the Phoenix Mine electrification tapped into the rhetoric surrounding resource potential and promised that electrification would enhance mining companies' ability to profit and meet colonial aims. One author wrote in the *Thames Adviser*, "The subjoined account of the application of electricity to gold mining purposes at the Phoenix Co. mine, upper Shotover, Otago, will be read with interest. It shows how the vast water power that annually runs to waste in, say a place like the Thames, might be made available for industrial purposes."[57] In the *Handbook of New Zealand Mines*, the author claimed: "The great want is capital. With that would come all the latest and best appliances for saving the gold, more systematic working, and the utilization of electricity and water-power on a large scale, similar to what has been done at Messrs. Bullen Bros.' mine [Phoenix Mine] at Skipper's. If English capitalists, who have invested millions of money [*sic*] in Indian, Californian, and South American mines, were to turn their attention towards the interior of Otago, they would get far better returns for their capital."[58]

National excitement about electrical power helped restore interest in mining, which had declined since the initial gold rush. Moreover, the promise of electric power for mechanical movement and lighting furthered British ideas about their technological superiority and prerogative to conquer the frontier and exploit its natural resources. An author in the *Telegraphic Journal and Electrical Review* said of the Phoenix Mine, "The importance of this new application of electric power for the lighting and complete working of an extensive mine like that of the Phoenix opens up a new era in the economic development of the very valuable and inexhaustible mineral resources."[59] The article explains the technologies involved, and offers praise for the inventiveness of the engineers. Still, it later concludes by arguing that the actual accomplishment of this project was "the means of transmitting power hither and thither—up a hill or down dale—down the mountain top or in the ravine below."[60] The generation of electric power by water meant that miners could transmit power to resource-rich quartz veins, use less water, be less dependent on consistent rainfall, and process more material.

Electricity appealed to investors, miners, and the colonial government because it promised increased efficiency, aligned with British ideas about technological progress, and improved water sources near mines. The promise of increased profit from the installation of modern electric technology allowed the press, owners, and investors to maintain a veneer of the utopian ideals of earlier settlement boosters. The electrification of the Phoenix Mine improved New Zealand's viability as a model settler-colony because it suggested modern power technologies could succeed across the colony, opening up the frontier for exploitation. Such sentiments connect these mines to commonplace colonial ideologies in Great Britain, expressed in the writings of Benjamin Kidd and Peter Lund Simmons, who believed the control and transformation of nature legitimized colonialism and capitalistic expansion.[61]

## ELECTRIC POWER OVER PEOPLES

The excitement around the electrification of Bullendale can obscure how the existence and productivity of the mine depended on the removal of Māori and the exploitation of Māori, settler, and immigrant labor. Despite British dreams of a New Zealand populated by a virtuous working class of British descent, efforts to establish a middle-class British workforce in the colony's new industries clashed with the mixed-race makeup of colonial mining operations. Workers performed arduous labor in the mines, with immigrants from different places living in makeshift housing across separate neighborhoods. While most of the miners came from Britain and Australia, many other Europeans, Māori, and Chinese immigrants also worked in Otago.

Because of their smaller numbers, Māori are often overlooked by historians in studies of the New Zealand gold rush, but they actively participated. West Coast papers frequently made note of Māori miners but seldom listed them by name, such as in this passage from *The Star*: "Chances of a rush of Māories [*sic*] have also to be considered, as . . . they have a keen appreciation of the value of gold, and will soon be on the spot."[62] Records produced by Māori miners during the Otago Rush remain scarce. Nonetheless, other miners and newspapers mentioned Māori mining exploits, such as a Māori from the 1860s, known as "Māori Jack."[63] Other Māori continued to appear in accounts and surveys of New Zealand's gold mines well into the twentieth century.

Chinese immigrants also represented a sizable percentage of the mining population.[64] In 1866 about 200 Chinese immigrants lived in Otago, but by 1871 their numbers had increased to around 4,300. Most came from the Guangdong Province in southeastern China, seeking wages and stability. Accurate census data for the population of mining towns is rare because of the fluid nature of the communities and colonial erasure of non-British settlers from the historical record. Many immigrants, particularly non-Europeans, avoided census takers to prevent deportation, taxation, and harassment.

A key, often overlooked connection between hydroelectricity and colonialization is that the transition to electric power took advantage of and even depended on Māori displacement from the land and transference of land ownership to settlers, to serve colonial economic gain and manufacturing goals.[65] Gold mining in New Zealand is itself a colonial practice.[66] The Otago Gold Rush catalyzed the invasion of a settler population intent on claiming land, mineral wealth, and water resources for personal gain. But even though the New Zealand gold rushes advertised the opportunity for settlement and individual wealth, the procurement of gold and the populating of New Zealand with British settlers was done primarily to enrich the empire. As a result, the colony moved closer to achieving the aims of the Colonial Office by establishing a primarily British population and expanding its disenfranchisement of Indigenous title and resources. The value of and right to obtain the resources in colonized land served as a central tenet of the "Doctrine of Discovery," a set of ideas and laws that sought to establish European cultural, religious, and cultural superiority to justify their acquisition of Indigenous lands. In New Zealand, this doctrine guided the installation of trading posts; the Treaty of Waitangi; and the subsequent exploitation of Indigenous peoples, resources, and land.

Without Māori cooperation and often exploitation, even on Cook's first voyage, European and British settlements would have fared poorly. The Māori provided explorers with basic supplies, guided traders to resources, forged strategic alliances, and became active participants in international commerce, especially in the timber, whaling, and flax industries.[67] This cooperation helps us understand that the Māori, along with the British, created a landscape desirable to British colonial interests and settlers seeking fortune, land, or a new life.[68] The Māori did not helplessly observe settler-colonization, nor were they just the New Zealand Wars' resistant noble warriors. Many general histories of the British Empire point to the Māori as one of the groups to most successfully challenge the British military. And while it is true the Māori were formidable in combat, we risk missing how the Māori also used negotiation.[69] Māori actively participated in colonization as victims and beneficiaries. Just as the British explored and attempted to understand how they might benefit from New Zealand, the Māori adjusted to the presence of the invaders and learned to navigate new economic and political structures.

In the 1780s, some Māori elected to travel to Europe and learn more of the world as sailors, missionaries, and artists.[70] Māori traders, already skilled in barter culture and becoming increasingly fluent in European languages, entered into more complex negotiations with Europeans. By the 1830s, many Māori had traveled worldwide, increasing Māori awareness of, and sometimes dependence on, European goods and imperial motivations. Trade with Māori remained central to the colonial economy leading up to 1840, but over time, exchanges increasingly exploited Māori interests. Settler trade and exploitation of resources whittled away Māori legal title to their land.[71] Before the European invasion, historians and demographers estimate that the Māori numbered around 100,000. During the first decades of the nineteenth century, the population shrank to an estimated 70,000–90,000 because of disease and warfare, especially the intra-Māori Musket Wars (1807–1842).[72] As the Māori population shrank, the settler numbers increased by the thousands, particularly after speculative mining reports appeared and the Colonial Office funded emigration to New Zealand. In 1840, New Zealand became an official colony under the Treaty of Waitangi, and British emigration increased dramatically.[73] With the post-treaty expansion of settlement, resource extraction increasingly depended on the removal of Māori agency, either through duplicitous diplomacy or outright violence.[74] By the time the Phoenix Mine switched on its electrical system, New Zealand's settler population had declined to 600,000.[75]

During the 1880s, hydropower continued to gain favor as settlers, miners, and merchants claimed previously occupied lands and water resources. By the 1890s, as hydropower systems became popular, other mines weighed taking further advantage of the water supply for electric power. Control over resources helped justify colonial actions and rationalize a mode of hydraulic consumption that guided the settler state's use of water for electric power into the next century.[76] The place of hydropower in the history of New Zealand's electrification cannot be reduced to geographic convenience, the availability of water, or some forward-looking ideals of environmental protection.

Hydropower efficiently provided electric power because of the settler-colonial infrastructures put in place around it.

New Zealand has 70 major river systems, with some 110,000 miles of river. The number of rivers in New Zealand has been the foundation of most of the country's history of hydroelectric projects. In one 1888 prediction, noting New Zealand's "innumerable swift-running torrents," the author wrote that "New Zealand will make more use of electricity proportionally than any other country in the world . . . not because the lack of fuels prevents extensive use of steam, for I believe her enormous coal fields will be practically untouched. There is no other land of any importance that is so well adapted for the production of electricity."[77] Today, most New Zealanders, with the exception of disenfranchised Māori, farmers, and environmental activists, take pride in the country's hydroelectric systems. They see hydropower as a method for keeping New Zealand energy independent and as an approach consistent with global green energy initiatives and climate change.[78]

The Phoenix Mine establishes the beginning of Aotearoa New Zealand's commitment to hydroelectricity. As a cornerstone of New Zealand's electrification history, the country's engineering community (past and present) celebrates its many technological adaptations as pioneering the hydroelectric systems that now generate an average of 57% of New Zealand's electric power.[79] Historians often herald the Phoenix Mine's electric system as technological genius, or as a foreshadowing of the country's hydroelectric potential. Some contemporary New Zealanders see the colony's utilization of hydroelectricity in the 1880s as a prehistory of environmental consciousness. Yet it is crucial to avoid uncritically attaching the Phoenix Mine's hydroelectric history to the country's current energy initiatives. Nineteenth-century electrification boosters did not concern themselves with green energy; environmental sustainability, if such a term can be applied at all, primarily concerned the utility of the land to the British Empire. Rather, the critical connection between early electricity transitions and recent projects is the notion of self-sufficiency. Making this connection helps us see the ways that colonial motivations and concerns materialized in New Zealand's electric grid.

The Phoenix Mine's electrification highlights how settlers made electrification align with settler-colonial interests, especially extraction. Electricity was used as a source of industrial power, not just for lighting or entertainment. New Zealand's mining engineers did not utilize electricity by coincidence or because it was easier. Instead, they believed electric power systems would increase New Zealand's potential as a resource frontier. They redefined the boundaries of that frontier by harnessing New Zealand's rivers and bringing technologies of the metropole into what had previously been nonsettler spaces. The application of electric power to the exploitative practice of mining demonstrates its colonial utility. Electric technologies like the stamper highlight the British commitment to their narrative of technological superiority and the creation of a model colony—self-sustaining, with a strong British settler population that would work to profit the empire. Furthermore, this mine establishes the beginning of New Zealand's commitment to hydroelectricity. Hydroelectric power did not always make more sense geographically or financially; rather, it appealed to those people vested in making New Zealand a more profitable, model colony.

CHAPTER 3

# THE STATION AT WORLD'S END

## SETTLER-COLONIALISM, THE DOMINION, AND THE LAKE COLERIDGE SCHEME

On September 26, 1907, the Colony of New Zealand became the Dominion of New Zealand. Congratulations from politicians in the form of telegrams sent to Joseph Ward, New Zealand's prime minister, rolled in from around the British Empire. William Kidston (1849–1919), premier of Queensland, Australia, wrote: "Trust New Zealand be as happy and prosperous as a dominion as she has been colony." Leander Starr Jameson (1853–1917), prime minister of Cape Colony, said: "On behalf of Cape Colony I desire to convey to you and the people of New Zealand our sincere congratulations upon the elevation of New Zealand to the status of dominion. We rejoice on your welfare and advancement." Joseph Carruthers (1857–1932), premier of New South Wales, pronounced, "I congratulate the people of your dominion on the event to-day. It is a recognition of the success which New Zealand has accomplished in colonization, and self-government."[1]

Celebrations happened around the islands, but none were as energetic as the events around the Parliament buildings in Wellington. The festivities at the new Dominion's capital included speeches, bands, and military demonstrations. One of the most visible and well-attend-

FIGURE 3.1. Dominion Day at the General Assembly Library Building. William Hall Raine. A.004858. Te Papa.

ed attractions was the lighting of the Parliamentary Library, where bright electric lights covered the building and formed the words "Advance New Zealand," "Colony 1840," and "Dominion 1907" (figure 3.1). Electric lighting displays had adorned public exhibitions, both industrial and political, at turn-of-the-century celebrations around the world.[2] Electric lights symbolized technological capability and modernity; in New Zealand, however, the relationship between progress and electricity had been associated with the settler-colonial mission to create a model colony for decades.[3]

The Dominion of New Zealand formally lasted from 1907 to 1947 and existed as part of the British Commonwealth of Nations. The category of *dominion*, deployed by the British Empire and certain settler-colonies between 1867 and 1948, defined a new sort of imperial relationship, that of self-governing colonies loyal but not subordinate to the Crown.[4] The use of the term *dominion* has a long history in Great Britain; colonies were the monarch's "dominions beyond the sea," whose governments belonged to the Crown. Dominions, such as New Zealand, Canada, and Australia, lived at the intersection of unity and independence. The first significant alteration to a British settler-colony's status by a colony that wanted to remain part of the empire happened within the Canadian Confederation, when Nova Scotia, New Brunswick, and the Province of Canada rewrote their constitutions, intending to become the "Kingdom of Canada." Some feared that title might evoke rebellious intent, so the new state was named the Dominion of Canada.[5] Canada became the model for the other "self-governing" colonies. The title elevated settler-colonies above subject colonies and prevented the adoption of potentially more rebellious titles like "kingdom."[6] This political transition offered greater political autonomy to settler-colonies in return for lowered imperial defense expenditures. The transition from colony to dominion represents the completion of the initial push for settlement because the colony achieved local self-government, one of the supposedly cherished aims of British colonialism.

The development of New Zealand's self-government accelerated at a rate unprecedented in the history of the British Empire. New Zealand underwent significant population shifts in the pre-Dominion years, from 2,000 Europeans and 70,000–80,000 Māori in 1840 to almost one million Europeans and approximately 40,000 Māori in 1907.[7] Even before the declaration of dominion, during the late decades of the nineteenth century, the New Zealand government rapidly altered its constitution to foster policies that granted Parliament increasing power while maintaining loyalty to the Crown. The legislative changes reflected colonial aspirations for the colony as an ideal England, first articulated by Edward Gibbon Wakefield, who said, "English society with its various gradations in due proportions, carrying out our laws, customs, associations, habits, manner, feelings—everything of England, in short, but the soil."[8] This idea was maintained and expanded by subsequent settlers, thinkers, and politicians.

Just as British politicians and writers celebrated the status of New Zealand as a model colony during the nineteenth century, so too did they portray the transition to dominion as the ideal conclusion of the settler-colonial mission. However, considering the political and cultural ties the country maintained with Great Britain before and after Dominion Day, New Zealand's transition should not be considered

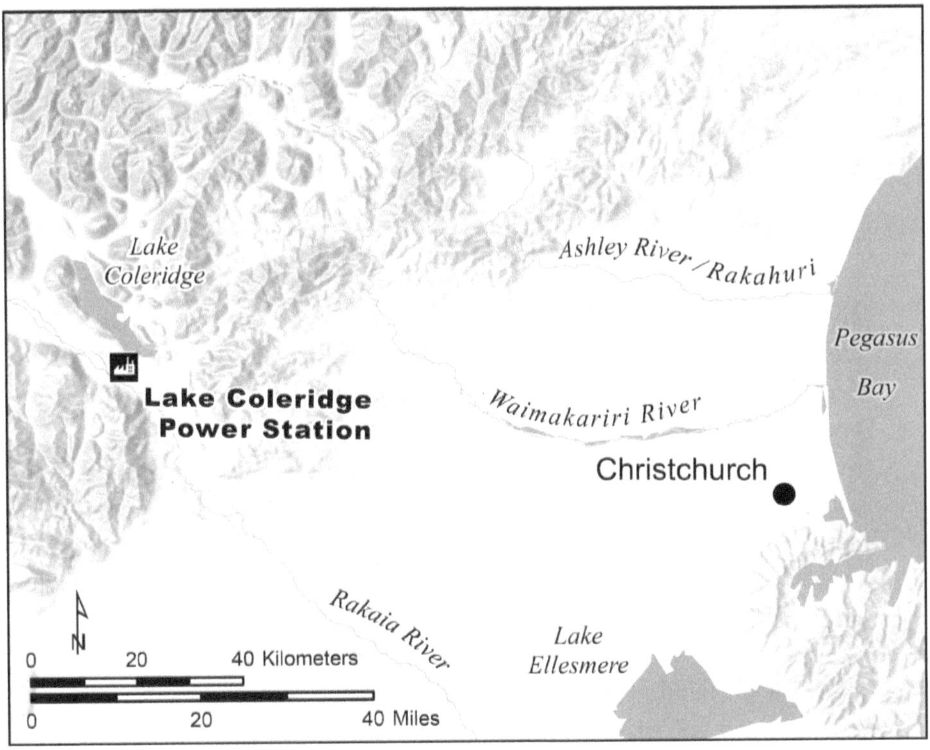


FIGURE 3.2. Map of Canterbury with the Lake Coleridge Power Station and boundaries of Christchurch.

an act of independence from the empire but rather a new manifestation of British colonialism. New Zealand remained part of the British Empire, despite the celebrations and speeches made that day about independence and separation from colonialism, and New Zealand government officials sought to maintain the authority they had enjoyed while under the control of the Colonial Office. The settler-colonial legacy remained intact.

Settler politicians believed that a prosperous New Zealand required the guidance of settler government, industry, and technology. Electric power bolstered New Zealand's seemingly contradictory plans for self-sufficiency on the one hand and economic productivity in service to the British Empire on the other. In the early years of the country's electrification, the most effective means of achieving economic gain and establishing colonial infrastructures, such as lighting frontier outposts, mining gold in the frontier, or building long-distance telegraphic communication, proved to be hydroelectricity.

The colony's government introduced new infrastructures, like state-run hydroelectric dams and generators, to more efficiently enforce control over water, land, and peoples. The proto-Dominion and Dominion government repurposed Indigenous land, established national electric power regulations, and supported the creation of a centralized Public Works Department (PWD) with the authority and resources to create hydroelectric infrastructures.

As New Zealand continued to favor self-rule policies, the PWD and its antecedents played a critical role in upholding the country's ideas about self-sufficiency and demonstrated its ability to operate independently. Parliament and the PWD wielded colonial rhetoric and practices to advocate for and produce state-run electric power, primarily where the control of land and energy resources was concerned. The establishment of the first nationally directed power station, which began with the Lake Coleridge Power Station in Canterbury on the South Island, exercised New Zealand's rebranded colonial authority under the auspices of dominion (figure 3.2).

The construction of the Lake Coleridge plant highlights how the New Zealand government effectively used decades of increasing authority for self-government while capitalizing on local desires to be productive and modern to begin building a nationally governed electric power infrastructure. Technological infrastructures allow states to establish an "apparatus of governmentality."[9] During the mid-nineteenth century, Colonial Office and PWD policies regulated the introduction of electric communication and power, increasingly granting authority to local political bodies. Nationalized electrification demonstrates the perpetuation of colonial infrastructures within New Zealand's new national government as they used and modified legislation concerning water resources and electric power to cement statewide control. The transition to electric power systems had fulfilled visions of a self-sufficient colony; the expansion and demand for electricity gave officials a means to facilitate a more centralized government that would continue to pursue a colonial agenda in New Zealand.

## BECOMING A DOMINION

New Zealand's transition to dominion represents the political reproduction of British settler-colonialism. Throughout the mid-1800s, amendments to the colony's constitution increasingly granted settler politicians authority to govern the island, which led to debates about how much self-rule the Colonial Office ought to allow. The first colonial constitution, the Charter of 1840, established rule under a lieutenant governor and an executive council. Together this group formed a Legislative Council and met to enact new laws in the colony. By 1846, municipal politicians and settlers had so frequently critiqued the Council's effectiveness that interest in establishing a new government arose around the colony.[10] After years of debate, in 1852 Governor George Grey introduced a new legislative structure to Westminster, which the Colonial Office accepted the following year.

The General Assembly first met in 1854 to establish ministerial responsibilities of the new body, sometimes referred to as a "responsible government," a system that embodies the principle of parliamentary responsibility, the British Parliament being a primary example. In 1855 the Colonial Office approved this transition to a responsible government and appointed a new governor. Governor Thomas Gore Browne (1807–1887) granted the General Assembly additional power but differed with them, primarily Premier Edward Stafford (1819–1901), on matters of "Imperial interests generally," referring to land and Māori affairs.[11] This agreement marked the beginning of responsible government in New Zealand under a dyarchy or "double government." Charles Hursthouse, author of *New Zealand or Zealandia: The Britain of the South,* and others strongly criticized what they considered an unbalanced and efficient distribution of power among the General Assembly, the Crown, and Colonial Office, an arrangement that politicians referred to as the "Treble Government." They wanted to remove the "quills" of Downing Street and give increased legislative authority to practice leaders in New Zealand.[12] Some critics, like Hursthouse, labeled the new parliamentary system as "double government" because colonial officers in London and settler politicians ineffectively shared power. Fraught communication between these figures frequently hampered decision-making. This delicate balancing act ultimately proved ineffective and served as an argument for increased self-government for the remainder of the century.

Ultimately, the so-called Māori Problem defined the limits of self-government because the New Zealand government lacked the means to support its military garrison. Journalists and politicians frequently referred to this problem from the 1860s forward. During that decade, it described what we now call the New Zealand Wars

(1845–1872) between settlers and Māori. After that, some authors evoked the phrase to refer to potential violence from Māori, Māori socioeconomic problems, and difficulties integrating Māori in New Zealand society.[13] In 1857, the Colonial Office had amended the constitution to grant basic administration of laws and domestic affairs to New Zealand's General Assembly, allowing it some semblance of self-governance. Still, contention remained over colonial and imperial authority in matters of colonial defense. Then, in the early 1860s, conflict with the Māori peaked and forced the issue of security upon local ministers. The Colonial Office decided that local governments should handle such matters, especially land redistribution.[14]

In 1861, Māori affairs became part of the ordinary business of New Zealand's Parliament. As a result of military difficulties in the Crimean War and the Indian Rebellion of 1857, Britain's Liberal government of the 1860s sought to change the size, organization, and expenditure of the British military. Under the direction of William Gladstone, Edward Cardwell (1813–1886), the secretary of war, led this movement, often labeled the Cardwell Reforms. The Childers Reforms followed these in the 1870s and 1880s, led by Secretary of War Hugh Childers (1827–1896).[15] Increasingly, the Colonial Office expected settler-colonies to protect their settlers in domestic conflicts, but the intensity of the New Zealand Wars proved an exception to this hands-off approach by the British Army.

The army's garrison in New Zealand had risen to 18,000 by 1864, but then the troops began to be redeployed, especially to Ireland to end rebellion and South Africa to protect mining operations.[16] The Colonial Office agreed to leave a battalion and £50,000 a year for "Māori purposes," an increasingly politico-racial position asserted by the government as they saw Indigenous peoples as a threat to their authority.[17] Funds for any further military action would need to be raised by the colony, which proved extremely difficult for the young administration. Although the "garrison controversy" concerned warfare, the debate proved formative to the settler push for self-rule and the local centralization of political authority. It remained a point of strain in the otherwise good relations between the colony and England for the remainder of the nineteenth century. Many colonists felt abandoned. As Henry Sewell (1807–1879), the first premier and third colonial secretary of New Zealand, argued, "it was not the treatment which a child would expect from its parent."[18]

The Gladstone government pushed for home rule, but this by no means meant an end to empire. Instead, he redefined what the character of the empire should be. During 1869 and 1870, Grey and Sewell led conferences of the self-governing colonies with the hopes of revitalizing trade and emigration. Politicians from Canada, Australia, and New Zealand met in the Cannon-Street Hotel to consider the relations between England and the Colonies.[19] The meetings did not alter any major Colonial Office policies toward the self-governing colonies. However, they played a crucial role in the new label of "imperialism" in the British Empire, which at this point meant forging a sense of unity across the empire. This thinking was best captured in an 1878 speech by Gladstone: "Let them flourish to the uttermost; and, if their highest welfare requires their severance, we prefer their amicable independence to their constrained subordination. The substance of the relationship lies, not in dispatches from Downing Street, but in the mutual affection, and the moral and social sympathies, which can only flourish between adult communities when they are on both sides free."[20] Although they only applied to settler-colonies' white male population, Gladstone's ideas about cooperation guided imperial policy well into the early 1900s. Even as the Colonial Office granted New Zealand more independence in self-government, the country remained culturally and financially tied to Britain.[21] By the turn of the twentieth century, New Zealanders still maintained strong cultural and familial connections in Great Britain. The connection became increasingly concrete as electric communication and faster naval transport shortened the gap between New Zealand and Great Britain. [22]

Despite the colony's increased economic success, many politicians, such as Premier William Fox (1812–1893), feared it would be vulnerable without imperial management of foreign affairs and defense. In response, politicians like Vogel pushed for New Zealand to act as an agent for British imperialism in the southern Pacific by extending his public works policies to consolidate the Polynesian islands as one dominion. He promoted the creation of a joint-stock company, known as the New Zealand and Polynesian Company.[23] As this idea met with little approval, Vogel opted for linking New Zealand with other colonies through federalization—a Pacific Federation. In 1883, the Confederation and Annexation Bill gave the government the power to link up with any Pacific island group, not under another regime.[24] Besides a slight movement to annex Fiji spearheaded by Richard Seddon (1845–1906), New Zealand's "sub-imperialism" only resulted in establishing a protectorate of the Cook Islands and Niue.[25] However, arguments for federalization with another Pacific island, Australia, were much more contentious. The colonies shared settler populations, trade, communication networks, and many shared interests, but New Zealand did not join the group that, in 1901, became the Commonwealth of Australia. In addition to these more locally concerned federalization initiatives, many New Zealanders were interested in an "imperial federation" or creating a single state among the colonies of the British Empire. This movement, led by the Imperial Federation League (1884–1893), never solidified and ultimately fell apart.[26] "The Premier was convinced that New Zealand's only hope lay in being designated in the World's Metropolis by some high-sounding, mouth-filling term," opined the *Feilding Star* in 1907. "So today, with the gracious consent of the King and his Ministers, our land has become a Dominion, and our people dominionites."[27]

## AN ELECTRIC STATE

For many New Zealanders, the change from colony to dominion was largely ceremonial and meant little for their daily life. One author in the *Evening Post* argued that "the man in the street has not quite realized the greatness thrust upon him, but he is enjoying the excitement."[28] Although many did not see much immediate change, increasingly centralized infrastructure ultimately would have real effects. With booms in economic and population growth, projects for improvement and expansion of infrastructure for the settler population drove national political agendas. This shift was driven by colonial treasurer Julius Vogel's (1835–1899) "Great Public Works" scheme, which borrowed some £10 million to fund road, railway, and communication infrastructure. The scheme ultimately borrowed a great deal more; by 1889 the amount rose to £37 million.[29]

Before this expenditure, New Zealand's infrastructure depended on local interest and short-term booms from extractive industries, such as gold, flax, and wool. The decades after the 1840 Treaty of Waitangi saw tumultuous changes on the islands. The country's diverse geography, the New Zealand Wars, and its poorly run provincial system virtually created two separate colonial spaces on the North and South Islands.[30] Settler violence, the forced relocation of Māori, and the New Zealand Wars ravaged the North Island, whereas the

South Island prospered from increased settlement, gold mining, and pastoral farming. But by 1870 the gold boom had slowed, long-term mining efforts replacing mass waves of panning, and the country's agricultural industry peaked.[31] Despite calls for a separation of the two islands into different states, most agreed that establishing a unified New Zealand was essential for the colony's survival.

New technological infrastructure projects provided a reason for the increased central government in the colony as more political power devolved from provinces to local authorities, a pattern that would reemerge when the dominion formed. In June 1870, Treasurer Vogel pushed the Immigration and Public Works Act through the General Assembly, which led to the foundation of the PWD a year later.[32] Over time the law enabled a system by which the New Zealand government would borrow large sums of money from Britain to subsidize immigration and fund railroads, telegraph lines, roads, public buildings, and port facilities. The success of Vogel's action led to the dissolution of the provinces in 1876 and a much stronger reliance on the national government.

In New Zealand's long history of electrification, the foundation for a national grid was provided by the legislation and practices concerning the telegraph. New Zealand's first law concerning electricity was the Electric Telegraph Act 1865, which gave the colony's general and provincial governments control over telegraphic communication. The act guided New Zealand's later laws regarding electrification and made the regulation, construction, and maintenance of electric communication equipment a central government function for decades. After the Public Works Act 1870, telegraphic legislation and construction remained under government control. Parliament, keen to assert its own authority, passed the Electric Lines Act 1884, which consolidated the preceding telegraph and telephone lines laws and created restrictions involving electric lighting for public places. The most significant amendments to the Electric Lines Act were the Municipal Corporations Act 1886 and its 1887 revisions, which granted local authorities the right to use water resources to generate electricity and sell it for lighting and motive power. Town councils, like the one in Reefton, or mining companies, like the Phoenix Mine, duly granted power boards the ability to harness water for producing electric power and to sell it. These legislative actions under the PWD guided New Zealand's national transition to electric power.

As in many other industrializing states, demand for electric power grew in New Zealand during the 1890s. Electrification promised to modernize the country's cities and manufacturing potential. Lighting remained the most popular application, but public transportation, department store elevators, and factory machinery also relied on electricity. Electric power markets continued to emerge, especially for mine owners and urban industrialists. Some cities located far from water sources adopted steam or suction gas systems,[33] but hydropower was the most preferred and widespread of the available options.

The success of the Phoenix Mine and Reefton inspired the development of numerous hydroelectric schemes in New Zealand, such as the Wellington City Corporation (1888), Stratford (1898), and Parihaka Pa (1899).[34] As with the two earlier projects, these schemes relied on municipal or provincial funding and, in turn, profit for the power providers. In 1896, with the passage of the Electrical Motive-Power Act, the pattern of electrification shifted toward national over provincial or municipal power boards. Richard Seddon, then the minister of public works, had pushed this bill through several readings in Parliament as a means of securing the use of natural waterfalls to generate electric power and better centralize the electrification of gold mining operations. Although some in the legislature pushed back against Seddon on the grounds that there might simply be too many waterfalls or that such an act was unnecessary, Seddon insisted that electrical energy production was the future and that this act would not grant the government too much power over land but rather would secure people's interests in using electric power.[35] The act stipulated that private enterprises could no longer gain control over any New Zealand waterway to produce hydroelectric power, nor could any local body grant permission to generate electric power without government consent. It effectively allowed the PWD to decide how to develop large-scale hydroelectric power stations and other electric power systems.[36]

Despite Seddon's concern for the interests of people who might use electric power generated by the country's natural waterfalls, it is worth noting that this bill gave the government an extraordinary range of control over the acquisition and control of land that might be used for electric power production. The bill and politicians like Seddon demonstrate a major legislative shift that provided the framework for larger power schemes to develop.

## UNLIMITED POWER, WATER, AND THE STATE

Exactly how plans for electrification formalized in the Canterbury region is difficult to trace, but by the 1880s the region's newspaper filled with announcements for electric lighting and power exhibitions. For example, in 1883, a passenger train carried two hundred people from Christchurch to the electric lighting display at Kaiapoi Woollen Mills, illuminated by two arcs and sixty incandescent lamps.[37] The event likely captivated many members of the Christchurch City Council and business community. Over the next decade, the council argued for public electric lighting and power supply, especially with the highly successful reports of electric systems in places like Reefton and the Phoenix Mine.[38] Electric power seemed to offer solutions for everything from revitalizing artesian wells to treating sewage.[39] Christchurch citizens and politicians hoped that increased electric supply would help the town and New Zealand more broadly to become a model state.

In addition to its general interest in electric power, the council advocated for hydropower. Most of the town's energy for lighting and electric power came from coal-fired steam,[40] and from the 1880s forward, the Christchurch press showcased the city's frustrations with early energy sources. Numerous viable water sources, such as the Waimakariri, Rakaia, and Opihi rivers, flowed in the Canterbury region around Christchurch. Orders for comprehensive surveys of the area for a hydroelectric scheme began in 1899, when the city council contracted Arthur Dudley Dobson (1841–1934), the city surveyor and engineer, to study the potential of Waimakariri River.[41] This survey marks the first phase of the planning that ultimately concluded in the construction of the Lake Coleridge Power Station.

Dobson's survey concluded that the Waimakariri River's width and shingle bed (a shifting gravel bottom) were unsuitable for a dam.[42] For Dobson and the other surveyors, constructing a power station in the region proved unfeasible, except for a single gorge. For many other rivers, including the Inangahua River, Dobson preferred to divert water from higher up the river through penstocks to a power station at a lower elevation. The report convinced members of the Christchurch City Council to gather information and cost estimates for such a power station. The Swiss firm Escher Wyss of Zurich, which had been responsible for dam development in Switzerland, Mexico, and Norway, offered to provide materials for £3,500 with additional expenses for the engineers and construction crews.[43] Some on the city council remained excited to begin construction on the scheme and hoped to start in October 1900, Canterbury's Jubilee year, because it was "naturally an occasion for erecting some tangible central memorial commemorative of the solid progress accomplished by Canterbury."[44] The construction of a large-scale public work would accomplish as much as the power supply because it would exemplify New Zealand as a modern, self-sufficient state.

Thus, even before the PWD decided to build a hydroelectric facility in the region, excitement among local politicians and residents mounted because the project represented progress for the city of Christchurch and New Zealand itself. Drawing on the earlier (and still very much alive) British settler-colonial mission, regional politicians stressed that electric power could reshape the city and nation into a modern manufacturing center, a prospect they hoped would excite the local population and gain its support. One of the most pronounced advocates for such a system was Thomas Edward Taylor (1862–1911), Christchurch Parliament member and mayor, who claimed that the Waimakariri River could make Christchurch the manufacturing capital of New Zealand. For politicians like Taylor, introducing electric power highlighted a half-century of "almost magical development" of the city of Christchurch and, indeed, New Zealand itself.[45]

Taylor produced numerous articles through Christchurch's newspapers, like the *Press* and the *Lyttleton Times*, concerning the industrial potential of Canterbury. For example, he published a long piece in the *Press* about how electric power produced at a large scale near Christchurch could transform the city into the manufacturing capital of New Zealand. In one section titled "The Colonies Are All Behind," he claimed, "New Zealanders are sometimes rash enough to boast of their reforming zeal and general progressiveness in comparison with the Old Country."[46] Here Taylor is chiding New Zealanders for being slow to align with British utilities that had allowed municipalities to begin competing with or taking over private companies. Despite the introduction of electric power systems in New Zealand, Taylor felt the country was lagging behind the United States, Great Britain, and other European countries in constructing electric power infrastructures. His remark is only partially true, especially among the British settler-colonies. Complaints of falling technologically behind were common almost everywhere in the late nineteenth century. Great Britain and the United States indeed had completed large-scale hydropower projects like Niagara Falls. However, such projects were not widespread, and the small-scale industrial stations standard in New Zealand during the 1890s resembled those of other countries.[47] South Africa did not begin major hydroelectric projects until 1905. Taylor admitted that Christchurch had "no Niagara" but had, according to Dobson, enough power to replace its current fossil-fueled infrastructure. At this point, advocates like Taylor and Dobson believed that hydropower could provide electric power more cheaply than natural gas or other sources.[48] In 1900, wrapped up in the excitement of the prospect of introducing electric power to the country, Taylor and a group of businessmen founded the New Zealand Electrical Construction Company. They intended to supply the city with equipment for an electric tramway system but ultimately failed to meet financial and legal requirements.[49]

As happened frequently with the regional governments in New Zealand, the Christchurch City Council needed help to effectively administer plans to build a large hydroelectric facility. In 1901, the council contracted Robert Hay (d. 1928), a civil engineer from Dunedin, to investigate the Waimakariri River.[50] He generally agreed with Dobson's earlier survey. Despite the promise of this site, the proposed scheme would significantly violate water rights in Selwyn County.[51] Regional farmers thought the construction would lead to blockages and washouts that could devastate the region's agricultural productivity. Ultimately, Hay recommended against the scheme. The Christchurch City Council and the Selwyn County Council did not reach an agreement regarding the project. The frustrations generated by these localized land and water disputes increasingly catalyzed the movement to centralize the implementation of electric power projects, giving responsibility solely to the PWD.

Although many remained convinced of the utility of the Waimakariri River, some were willing to seek alternative options. In 1901, Taylor compiled another regional survey, including Dobson, city council members, and journalists, of Lake Coleridge and the Rakaia Gorge.[52] Electric power proponents like Taylor thought that a power station at Lake Coleridge might prove just as challenging to build but that the outcome would pay for the hardship. Taylor meanwhile continued to advocate for the Waimakariri River scheme. The power station mattered more than the location—so long as it was near Christchurch. Taylor's excitement kept the scheme alive despite the hesitation of the Christchurch City Council. After traveling through Europe touring power stations, Taylor proposed that the city hire a Swiss engineer, Colonel Théodore Turrettini (1845–1916), but the city council ignored his proposal. Turrettini briefly worked for Edison in the 1880s but returned to Geneva to work on the hydroelectric power stations there.[53]

The debate continued into 1902, especially in the aftermath of the proposition of the City of Christchurch Electric Power and Loan Empowering Bill. Christchurch Member of Parliament William Whitehouse Collins (1853–1923) introduced the bill to empower the city to control and seek financial assistance with electric lighting and power. It sparked debates over financial responsibility for the Waimakariri scheme. Taxpayers in Christchurch worried that they would pay for everything while the entire region benefited. Additionally, Selwyn County and the Waimakariri-Ashley Water Supply Board still feared a loss of irrigation supply. The city council eventually added a clause that protected the river within a certain distance of the gorge.[54] On November 17, 1902, the bill successfully passed through Parliament, authorizing the city to raise a loan of no more than £250,000 to contract, construct, and maintain a power station on the Waimakariri River.[55] However, too many protested the scheme, especially when many engineers had reported that the Lake Coleridge proposal was less dangerous.[56]

The city council continued to hesitate in finalizing plans for the power station until the passage of the Water-Power Act 1903 complicated their decision. The act stated, "The sole right to use water in lakes, falls, and rivers or streams to generate or store electricity or other power shall vest in His Majesty."[57] By making water a resource exclusively controlled by the national government, this act effectively placed electric power under its control, because water generated most of New Zealand's electric power. For the first time, the national government regulated almost all electric power facilities. Parliament's growing legislation to control the country's vast water resources represents one of the most vital links between colonialism and hydroelectrification in New Zealand, as politicians directed water usage for electric power to ensure the national government's control over settler and Māori energy interests.

Although the state permitted the city to construct the Waimakariri power station, the national government's new authority over water resources led the Christchurch City Council to hold back. Seeking to maximize the country's hydroelectric potential, the New Zealand government contracted L. M. Hancock, an American electrical engineer, and P. S. Hay, a civil engineering superintendent from the PWD, to report on potential sites. They surveyed the entire country. In 1904, Hancock published his report and submitted it to Parliament. He assured the government that the country's water power was "ample for all existing needs and any future growth and extensions that may ever be devised."[58] As to Canterbury, Hancock reported, much to the dismay of citizens of Christchurch, that the Waimakariri "was best left to its own sweet will."[59] However, he suggested that Lake Coleridge showed promise, calculating that it would yield 94,677 horsepower, whereas Waimakariri promised only 20,440. Furthermore, he argued that the area around Lake Coleridge had room for expansion and workable terrain.[60] Council members and advocates for the scheme complained that Hancock was an expensive foreigner who could not give the region fair treatment, while Taylor said that Hancock's report "was not worth the paper it was written on."[61] Nevertheless, the Hancock Report made it clear that Lake Coleridge, not the Waimakariri River, should be the national government's priority, and the city council begrudgingly relented.

For consumers and the politicians who advocated for it, affordability drove interest in building a Lake Coleridge Power Station. They believed that the cheaper energy promised by the council and engineers would turn Christchurch into a modern manufacturing city. In almost every positive article about the Lake Coleridge Power Station, authors emphasized the cost benefits of electric power. It would be cheaper than natural gas and become more affordable as more subscribers took advantage. In 1904, ahead of the construction of the power station, the *Evening Post* reported, "More favourably than any other water supply [...]the Christchurch and tributary districts could be very effectively supplied, that the installation would be simple and that the initial installation would not be enormously costly, as the plant could be enlarged as needed."[62] This statement is one of many excited reports publicizing the potential of a Christchurch-serving hydroelectric power station on the Rakaia River watershed, which included Lake Coleridge in the early twentieth century.

However, the proposed water diversions and scale of a scheme at Lake Coleridge, which investors projected could cost £1.2 million, initially discouraged the PWD from seriously considering the plan. In the first years of the twentieth century, Public Works funding focused primarily on transportation, connecting the islands by increasing the flow of people and materials. Many PWD officials believed that trade and transportation would produce unity among the islands' citizens and municipalities, particularly completing the North Island Main Trunk (1908), a 680 km railway system.[63] Electric power did not receive as much attention because of the previous commitment to connecting the islands by road and rail.

Ultimately, electric power boosters succeeded by promoting its manufacturing potential. With Christchurch as a distribution center, the Canterbury region functioned as the South Island's major agricultural producer of seed, wheat, wool, and meat. Since introducing refrigeration technology, companies such as the Canterbury Frozen Meat and Dairy Produce Export Company represented large Christchurch and New Zealand economic sectors.[64] Additionally, the town hosted numerous port-related businesses, metal processing factories, and engineering firms. At the turn of the century, Christchurch politicians competed with other cities, such as Wellington and Auckland, to expand their importance and attract a large labor pool. Increasingly, city residents and politicians insisted they needed electricity to keep and attract businesses to Christchurch.

By 1909, the Christchurch City Council reasoned that the town's existing steam plants could not meet the growing demand for electric power. The New Zealand government had not broadened its regulation of electric power since 1903 because of the Water-Power Act. In 1910 the Christchurch *Star* opined, "If the government installed its power schemes, the profits would go into the consolidated revenue, and would be used to the advantage of the whole dominion . . . The proper use of water-power would prove to be a source of enormous profit in the near future, and that it would probably rival the railways in regard to the production of wealth."[65]

As such an expansion would serve Christchurch's manufacturing businesses and rising electric power usage for lighting, the city council argued that either the town or the national government should fund a significant hydroelectric facility to keep New Zealand's (or Christchurch's) economy expanding. They quickly reconsidered the Waimakariri River scheme and dismissed it because of the same water rights concerns.[66] Considering all possible alternatives, they even unsuccessfully petitioned to build a city-run power station on Lake Coleridge. After the petition failed, Mayor Charles Allison (1845–1920) jump-started a campaign to enlist government support—specifically the minister of public works, Roderick McKenzie (1852–1934)—to fund a power station at Lake Coleridge. He claimed that "without electricity, Christchurch will lose its status as a manufacturing city."[67] In response to Allison, McKenzie felt that the government was obligated to supply electric power to the city but believed the expense prohibitive. Financial gain and the potential of expanded manufacturing capabilities continued to drive Christchurch's and Parliament's interest in constructing a state-funded and regulated power station. Finally, Lawrence Birks (1874–1924), an Australian electrical engineer and lead engineer of the Tourist Department in Rotorua, submitted a report on the hydroelectric potential of the Lake Coleridge and Hutt River schemes to the New Zealand government.[68] Another report written by R. W. Holmes, chief engineer of the PWD, supported Birks and came to similar conclusions about Lake Coleridge's benefits, much as Hancock's report had done. After years of resisting, Taylor agreed to drop his promotion of the Waimakariri River scheme in favor of Lake Coleridge.[69]

Not all New Zealanders maintained so much optimism about the development of a power station at Lake Coleridge, however. For example, Frederick Black, an electrical engineer, harshly criticized the seemingly carefree commitment to the scheme in the press:

> Nobody knows what they [?] will ultimately cost, nobody knows what will be done with them [the power stations], and nobody—or almost nobody cares. The main thing is that nearly two and a quarter million sovereigns are down on the list to be spent. The Government say and the public believe, without knowing why they believe, that the pouring of this golden stream into New Zealand rivers will create industrial conditions in the land at present undreamt of, that a new era will dawn wherein electricity—blessed, mysterious force—will permeate all our lives, and that, by virtue of the great water gift of the gods we shall soon tower above all the nations in the Pacific.[70]

Despite critics such as Black airing financial concerns in the press and handwringing in Parliament about the cost, on November 4, 1910,

FIGURE 3.3. Traction engines at Coalgate railway station. *The Press* (newspaper): Negatives. Ref: 1/1-008122-G, Alexander Turnbull National Library, Wellington, New Zealand, https://natlib.govt.nz/records/22393432

the official Public Works study began. Even though they initially feared the high cost, the government passed the Aid to Water-Power Works Act 1910 to credit the Public Works Fund for hydroelectric power and irrigation.[71] McKenzie did not explain the change in Public Works policy explicitly. However, local papers argued that the success of other major hydroelectric projects in America, Canada, and Switzerland had supported impressive industrial results and suggested the same would be true for New Zealand.[72]

Construction projects followed the next year. Before the building could begin, the ground required testing for strength, materials for waterproofing, and a surge chamber for rerouting overflow. The supervising engineer, F. T. M. Kissel, began constructing a "town" for the laborers and future workers for the power station, even moving his family there. Workers eventually labeled the town Kisselton, and locals did not appreciate the name coming from outsiders. One of the inhabitants thought the locals—especially the Murchisons, who owned much of the sheep-grazing land in Canterbury and farmed much of the region during the late nineteenth and early twentieth centuries—ought to have a say.[73]

The town at first consisted of eight cottages built to withstand the fierce wind and rain. Word spread that work was available, and laborers flocked to Lake Coleridge, where they made progress on the tunnels, buildings, and pipelines. Builders constructed two tram lines to transport workers up the Rakaia River and remove soil. In continuing with the PWD's transportation mission, engineer F. W. Furkert (1876–1949) surveyed and built roads between the station and Coalgate, the nearest railway head, so that tractors, "Furkert's Motor Cars," could quickly move material to the station. Tractors, traction engines, and horse-drawn wagons moved material between 1910 and 1914 (figure 3.3). They transported 1,750 ironbark piles to support lengthy 66 kV transmission line, 500 wooden poles for local distribution in Christchurch, 130 miles of copper 7/14 Standard Wire Gauge (SWG), 60 miles of copper 7/12 SWG, 420 miles of 7/.135 aluminum wire, 14,080 yards of insulated cable for the 11 kV underground lines, 12 General Electric transformers for the powerhouse and substation, and three Escher Wyss Francis-type 1,500 kW generators.[74] Using a combination of rail, tractor, and animal power, the workers transported some 12,000 tons of material in total.[75]

Although earlier portrayals of the area as perfect for a large power station excited the city council and PWD, the miserable working conditions, unknown geological barriers, and disgruntled workers proved otherwise.[76] The environment, ever ignored by the architects invested in these electrical imaginaries, challenged the commitment of Public Works and Christchurch officials. The geological difficulties alone tested the resolve of everyone involved. The power station was the first to be constructed on a glacial moraine. Its loose gravel made construction difficult because it required a deep concrete foundation, a temporary solution that plagued the power station throughout the twentieth century.[77] Work frequently stalled due to harsh weather. Miscommunications caused delays for Kissel and his crew throughout the construction process. One of the workers, James Ryan, claimed that "these were the worst conditions he had ever known."[78] Besides the working conditions, Ryan and other workers interviewed for an article expressed their discontent with the living conditions. Part of their discontent stemmed from the unequal treatment of laborers and other PWD members, the latter of whom got to live more comfortably. According to the laborers, the rate for food and board was outrageously high.[79] Other workers took issue with the safety of the equipment and speed of construction. As a result, the Canterbury General Laborers' Union formed and renegotiated wages between contracted companies and Public Works.

The PWD maintained that only a central governing body could handle major infrastructure projects, so the agency became increasingly involved in construction. As had been the case with many colonial projects before, New Zealand's terrain required new designs, so electric power boosters argued that centralized authority and modern technology would fix the problems. After all, it had worked before, when the Colonial Office had directed finances toward expanding mines to the South Island's West Coast or exploiting the tourist industry potential of the hot springs on the North Island. In 1913, under the direction of Lawrence Birks, the PWD took over the work to make regulation easier. Birks appointed Evan Parry (1865–1938) the lead electrical engineer. Parry was one of Lord Kelvin's former students at Glasgow—a reminder of how this local electrical project remained connected to global electrical science and engineering networks of the period. Many supporters believed construction would stop halting, saying things like all work "is likely to go better now."[80] In effect, the boosters of this project, and of electrification infrastructure more broadly, believed that the national government would make the smaller oversights that plagued the early construction disappear. They did not. During the last years of construction, strikes in England delayed the shipping of turbines and generators, and an exploded safe in the laborers' lodgings and a labor dispute with the contractor in charge of digging the tunnel all delayed the completion of the power station.[81]

Workers finished the erection of posts and wires through most of the eighty-five-mile distance to Christchurch, where citizens eagerly awaited increased access to lighting and manufacturing jobs. By July 1914, the six major sections of the scheme—the tunnel, pipeline, powerhouse, and transmission lines, a substation in Addington, and the Christchurch City Council's plant—had been completed.[82] The power station generated 66,000 volts, which were transformed at Addington to 11,400 volts for distribution. From 1913 forward, once the construction of the power scheme seemed a reality to the project's advocates and future customers, papers frequently printed excited statements about the affordability of electricity to be produced at Lake Coleridge: "Under the present rates electricity in the house has been found decidedly cheap, and when the Lake Coleridge scheme comes into force the expense will be much less."[83] As construction ended, negotiations concerning the cost of electricity demonstrated that the Public Works office and city council were willing to forgive prior difficulties because of the economic potential of hydropower.[84] Earlier in 1914, the PWD and the Christchurch City Council agreed that the city would take a minimum of 500 kilowatts, and the power station agreed to supply 4,500 kilowatts. Escher Wyss, the Swiss firm, provided the three 1,500-kilowatt generators. The council agreed to pay the PWD £8 per kilowatt for the first 300 kilowatts and £5 for every kilowatt beyond those in quarterly installments, which is how Christchurch and other municipalities charged the patrons purchasing their power.[85] Years later, Birks used this cost as an example of the economic efficiency of hydroelectricity when he classified it as "poor man's light."[86]

Even though the city council eventually agreed to these rates, they feared that local companies might try to circumvent the city and purchase power directly from the government. To address these concerns, the council and the minister of public works, William Fraser (1840–1923), arranged a meeting on March 5, 1914. There, Fraser stated, "The government had sunk a lot of money in the Lake Coleridge Scheme and were anxious for it to be a commercial success, but it would not make both a wholesale and retail profit."[87] However, he also acknowledged that while the government did not wish to compete with the city council, local bodies could not control the government if, in the future, the PWD elected to negotiate with other parties.

Lake Coleridge began supplying power to nearby Lyttleton Harbor and Tai Tapu soon after Christchurch. Excitement swept the region as more people waited, sometimes impatiently, to connect to power. After 1915, the city and surrounding boroughs bought into the power scheme, and the Public Works Annual Report stated, "The plant has worked smoothly and efficiently, without serious interruption, and the country will learn with satisfaction that this important commercial venture on the part of the Government shows every promise of becoming a financial as well as an engineering success."[88]

The new power station meant more than an increased electricity supply in the region. After its completion, settler politicians and consumers heralded it as a turning point for the Dominion, echoing the rhetoric of self-sufficiency used to bolster support for Reefton and the Phoenix Mine. As the author of the *Lyttleton Times* announcement exclaimed, "The important fact is that with the practical completion of the Lake Coleridge installation New Zealand is fairly entered upon an undertaking that is going to benefit Canterbury enormously in the immediate future, and that may revolutionise the industries of the Dominion with the next quarter century."[89] In a speech during the opening ceremony, Prime Minister Heaton Rhodes (1861–1956) proclaimed "that in time to come the whole of the power required in the Dominion would be generated by the current generated in the streams and rivers of the country."[90] Furthermore, observers celebrated New Zealand's "endless supply," which had, thanks to the political maneuvering a decade earlier, under people like Richard Seddon, come under the control of the state. "The late Mr. Seddon, with his far-seeing statesmanship made the abundant water power of New Zealand a State monopoly, and when the turbines begin to spin at Lake Coleridge the people of this country will appreciate the significance of that step to a land that possesses practically unlimited supplies of the 'white coal.'"[91]

## THE TRIUMPH OF HYDROIMPERIALISM

Understanding Lake Coleridge's place in local enactments of colonial power, such as its adherence to the Colonial Office's emphasis on centralized governance, is aided by placing the project in the context of the more significant global transition to hydroelectric power, especially in British colonies and territories. Dams and other hydroelectric systems, such as the Lake Coleridge Power Station, existed within an imperialist dynamic in which settlers imagined the rivers and lakes as an industrial resource. As we have already seen in New Zealand, settlers altered the environment around the Inangahua River in Reefton and Skipper's Creek at the Phoenix Mine to create a sustainable settlement and enhance economic productivity in the service of colonial interests. Historians of electrification often take the widespread geographic availability of water for granted and suggest that hydroelectric power must have been the most logical, apparent, or straightforward course of action for stakeholders interested in electric power systems. But we ought to undo such assumptions. No one energy system is the obvious or natural choice. Rather, they are produced by the environment, the people who live there, and the people who control its resources. In colonial spaces, water manipulation and regulation reveal and reproduce unequal power regimes. Across the empire, the British enacted water regulations to control settler populations and marginalize Indigenous peoples.[92] Historians and sociologists classify this type of technopolitics—the use and management of water resources in a colonial setting—as *hydroimperialism*.[93] Placing the construction of Lake Coleridge in the context of hydroimperialism situates New Zealand's transition to electric power within the broader frame of imperial utilization of water to centralize national government control over Indigenous and settler populations in colonized places. The Victorian scholar John Miller Dow Meiklejohn (1830–1902) argued in one of his geography textbooks that "Divine Providence, brought it about that the people of Great Britain are now the traders and news carriers for the whole world. These functions have given us another office, have forced upon us another mission. This is to keep the Waterways of the world—the water-ways in ocean, sea, lake, river, and canal. Great Britain is, therefore, the Guardian of Water-ways."[94] The success of hydroelectric schemes had to do with geographic availability, but the manipulation of water also facilitated hydroimperialistic ideas about making the country's natural resources profitable.

In the 1840s, to revitalize urban spaces, Great Britain undertook sweeping water supply and drainage reforms at home and reproduced those changes in their colonies. Besides hydroelectric systems, many colonial cities, like Bombay (Mumbai); Colombo, Ceylon (Sri Lanka); Hong Kong; and Singapore, installed massive water schemes to solve labor, health, and environmental problems.[95] British officials and engineers hoped that new water systems would support the productivity and modernization of the colonies. At the turn of the twentieth century, hydroelectric infrastructures and the accompanying political institutions (like New Zealand's PWD) expanded rapidly in the British Empire, especially in Canada.[96] The empire used water rights to expand influence in numerous colonies such as India, Canada, and those in Africa for drinking water and hydroelectric power. Colonizers across the British Empire and in other settler nations, like the United States, enacted water regulations to control settler populations, marginalize Indigenous peoples, and justify new infrastructures—see, for example, the formation of the Bureau of Land Reclamation in the United

FIGURE 3.4. Electricity generation at Lake Coleridge Power Station, with insulators on the roof of the turbine station. Photo undated. Samuel Heath Head Collection. Ref: 1/1-007272-G, Alexander Turnbull National Library, Wellington, New Zealand. https://natlib.govt.nz/records/22517881

States (1902) or the signing of Treaty No. 9 in Ontario (1906).[97] The construction of new colonial water supplies increasingly transformed previously decentralized systems into centralized schemes with new restrictions. While the power boards operated locally, water rights had to be obtained from the colonial government.

Lake Coleridge's construction depended on imperial power dynamics and exploitation. While the Water-Power Act 1903 proved a major inconvenience for settlers and Māori farmers dependent on rivers for irrigation, the removal of water rights from the Māori was especially harsh because of the central role of water to Māori cosmogony and community life. The PWD and Dominion government's efforts to control water and build power stations must be understood as part of a broader settler-colonial project.[98] Rigid control over water facilitated the legislative and eventually the physical structure of the settler vision for New Zealand. Hydropower and the water-controlling bureaucracies of the British Empire became another settler tool for trying to create an ideal New Zealand that excluded Māori culture. Water remains a contested Māori-rights issue today.[99]

The Public Works Amendment Act 1908 made it possible for the New Zealand government to purchase the Lake Coleridge Power Station land. An amendment to the first Public Works Lands Act in 1864, the 1908 act granted the government ill-defined powers to acquire land. As the law allowed for a lower compensation rate, it enabled disproportionately higher rates of acquisition of Māori land. For example, on November 25, 1914, the same day that the prime minister of New Zealand, William F. Massey (1856–1925), delivered a speech to commemorate the opening of the Lake Coleridge Power Station (figure 3.4), the Waiau Native Reserve submitted a petition requesting that the government return 1,000 acres on the west bank of the Waiau River, about 70 miles north of Christchurch. Due to clerical oversight, neglect, and possibly contempt, the government had taken the land in the 1860s, disregarding treaty protocols. The Public Works Act had further legitimized, at least in the eyes of the settler government, the acquisition of the land, transforming it into an official piece of the new Dominion's modern utility infrastructure.[100]

Despite the tensions over the initial investments in surveying and material between the City of Christchurch and the Public Works Department, Lake Coleridge, at least for public works officials, proved the effectiveness of the PWD and national government in constructing and operating a hydroelectric scheme. The Lake Coleridge scheme harnessed hydropower on a larger scale and at a higher cost than any earlier electric project. Over the next few decades, the station added more generators and expanded its supply network. Later in the twentieth century, the PWD fed new rivers into the lake to increase output, replaced generators, and connected them to other power boards. Today the station remains an essential plant in the TrustPower network (now Mercury NZ Limited) on the South Island.[101] Historians of electric power in New Zealand, such as Neal Rennie and Helen Reilly, credit the scheme as the origin and earliest example of New Zealand's hydroelectric capabilities, setting forth a model for subsequent electric power infrastructures.[102] Christchurch became the primary focus in New Zealand for implementing the state-funded electrification that came to dominate the expansion of electric systems for the next century. The infrastructures established by the PWD, especially electric power and communication networks, provided the physical framework that facilitated more centralized authority within New Zealand. As power demand increased, the national government continually found regulation effective in maintaining and expanding its influence. As others have noted, governments distance themselves from infrastructures to control and organize populations through technological domains that seem separate. By managing infrastructures like electric power systems in New Zealand, its Parliament played an active role in municipal boundaries and regional industries.[103]

The Dominion movement, the expansion of the PWD, and the orchestration of the hydroelectric works at Lake Coleridge reinforced British settler-colonialism and underlined the imperial practices embedded in New Zealand's electric power infrastructure. Much like other late-nineteenth-century technological infrastructures such as the telegraph and railroads, electric power systems and the bureaucracies that crafted them justified the increasing number of calls for local governance. As scholars like Frank Edward and Mikael Hård have argued, "The British Empire was based on a firm material foundation. Harbour facilities, railroad tracks, military equipment, telegraph lines, bitumen roads, diesel generators, water pipes and sewage culverts were necessary to control the space where the sun never set."[104] The new Dominion government recognized this and orchestrated the Lake Coleridge Power Station to maintain power. New Zealand, perhaps most among the settler-colonies, strived to achieve self-sufficiency at the end of the nineteenth century. A large-scale nationally funded electric grid promised to increase the productive capabilities of Christchurch and subsidiary communities. Furthermore, it demonstrated to New Zealanders and Great Britain the young Dominion's ability to complete national-scale projects. However, Lake Coleridge was not an outcome of New Zealand's newly gained independence but rather the product of more than a half-century of colonial policies aimed at creating a self-sustaining and productive British population in the Pacific.

# CHAPTER 4

# "THE MEGAWATTASAURUS"

## THE HUNTLY POWER STATION AND NEW ZEALAND'S ENERGY REGIME

> The Public Works Act will not be used to take Māori land.
>
> **—Minister for Public Works Hugh Watt, 1973**

After the 1914 opening of the Lake Coleridge Power Station, the Hydroelectric Branch of the Public Works Department (PWD) declared it a success. There was even a weekly meeting in Christchurch where Lawrence Birks, the electrical engineer in charge of the project, explained how the power station would supply power to people "not in the usual technical way."[1] He described, using illustrations and nontechnical language, how power lines distributed electric power to drive trams, illuminate the streets, and light up shops. Christchurch began receiving power twenty-four hours a day in March 1915, and the surrounding region, namely Lyttleton Harbor and Tai Tapu District, followed suit.

The Lake Coleridge Power Station and the expanding grid experienced problems, but the project launched Aotearoa New Zealand's electrification into an era of rapid growth, especially after World War I. Broad support and demand for electricity existed among the settler population. Advertisement columns of New Zealand's newspapers

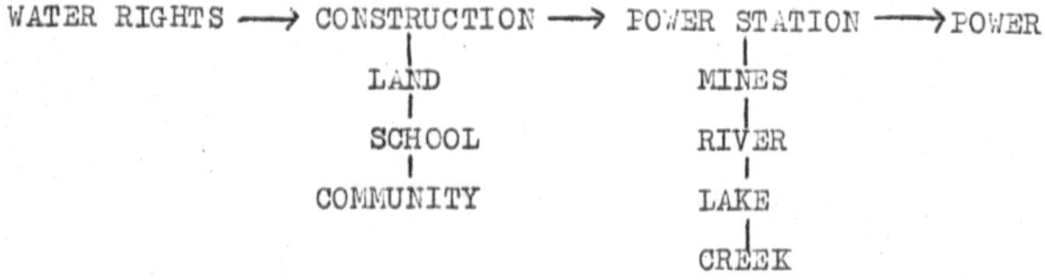


FIGURE 4.1. R. Mahuta, "Planning Issues and the Huntly Power Station" (1978)

during the first decades of the twentieth century frequently featured electrical domestic and industrial equipment sales. The National Library maintains a marvelous collection of oral histories and ephemera showcasing this transition.[2] Outside of those ads, newspaper forums and minutes from town hall meetings showcased debates about electricity rates. Citizens vented frustrations at the national government and local power boards, but the demand and perceived need for electric power remained strong.

Following a deluge of hydroelectric expansion during the 1920s–1940s, the New Zealand Electricity Department (NZED) formed in 1958. During the 1960s–1970s, the department built numerous thermal stations, the most well known of which was the Huntly Power Station (constructed 1973–1983). New Zealand's largest thermal power station to date, Huntly, whose steam turbines could burn natural gas or coal, was built amid New Zealand's broader environmental movement, a widespread national closure of coal mines, and growing Māori advocacy for land reclamation and environmental protection. During this period, environmental advocacy groups, new government agencies (like the Ministry of the Environment), and Māori communities in and around Huntly (such as the Tainui), pressured the state to assess the environmental impact of infrastructure projects. As public works officials increasingly saw thermal generation as a solution to power woes and economic instability, these groups actively pushed back. One of the critics of Huntly's construction, Robert Te Kotahi Mahuta (1939–2001), a prominent Tainui leader from the Waikato region, challenged the construction process for a decade. He created the diagram featured in figure 4.1 and highlights a simple but potent throughline connecting the Huntly water-rights legislation to nineteenth-century electrification—power over people and land—in service to the aims set forth by nineteenth-century colonial authorities and the policies they created.[3]

This chapter unearths the colonial rhetoric and intent in the origins and execution of the plans to build Huntly Power Station. Settler identity in New Zealand, as noted in previous chapters, had long been tied to ideas of self-sufficiency. In World Wars I and II it also became associated with resilience, both real and imagined.[4] These attitudes framed much of the support for hydropower and the extractive industries of the preceding centuries. How that language and the acts of dispossession featured in the Huntly Power Station's planning is worth studying, not because the project differs from earlier colonial or Dominion initiatives but because of how long-familiar ideas about resources, Indigenous peoples, and infrastructures remained in postwar New Zealand. The Huntly project is a product of the colonial systems in New Zealand that violently stole land, resources, and life from Māori in Waikato. Although the Ministry of Works (MoW, which officially absorbed the PWD in 1948)[5] deployed new policies and arguments about energy security, many of the same policies and bureaucracies that promoted electric power in nineteenth-century mines or at Lake Coleridge played into Huntly's construction.[6]

Beginning in the late 1960s, NZED mirrored colonial-era actions by manipulating environmental regulation, stripping people of water rights, and continuing to confiscate land from the Tainui in order to produce energy. The primary function of New Zealand's electric power infrastructure, even during the late twentieth century, remained its ability to protect the settler-colonial vision of a self-sufficient Anglo-dominated state in the Pacific. The example of the Huntly Power Station works as a stark reminder of the colonial roots of New Zealand's electric infrastructure in which the PWD/MoW reconfigured policy to suit their ideas about the country's energy infrastructure.

## ELECTRIC POWER AFTER LAKE COLERIDGE

Beginning with Lake Coleridge, the PWD continued to argue for the centralization of electric power production. They hoped to accomplish this by enlarging the country's power production capabilities to fuel New Zealand's productivity and expand the government's control over its energy infrastructure, primarily to extend it to growing urban spaces like Wellington and Auckland and support productive industries, like dairy and meatpacking, that depended on electric power.

In 1918, Evan Parry, chief electrical engineer of the PWD, prepared plans for the country's electric grid and modeled it on Canada, Sweden, and Australia, where a central government agency controlled electric power systems. He wanted to facilitate the state's responsibility to manage public hydroelectric supply rather than adopt the private enterprise system like Great Britain or the United States.[7] The administration argued, just as it had with Lake Coleridge, that only the national government, acting through local power boards, had the means to build the necessary infrastructures—at least those suited for industrialization and settlers' vision for New Zealand.[8]

In his 1918 report to Parliament, Parry argued for the expansion of hydroelectric schemes to power cities on the North Island:

> The development of roads and railways will call for urgent attention, but it seems likely that the chief part of the public works programme of the government will be the utilisation of water powers for the production of electric energy. . . . The hydroelectric works at Lake Coleridge have been so successful that not only is the Government convinced by their success of the practicability of developing other schemes, but the general public is prepared to press the authorities (and is already pressing them) to proceed with the large programme that has been drawn up. The programme provides for a series of schemes designed to supply the whole of the North Island.[9]

The program to which Parry refers is a blueprint that sketched out the construction of the Mangahao, Waikaremona, and Arapuni hydroelectric schemes. In 1924, the PWD completed the Mangahao Power Station, which provided electric power to North Island cities like Wellington and counties populated primarily by settlers. The dam also encouraged interest in electric power in cities like Auckland and Wellington, which followed the pattern of population growth in New Zealand, where the North Island population continued to soar throughout the twentieth century. Other major hydroelectric facilities like the Arapuni Power Station (1929), which would supply much of the power for Auckland, were built during this period.

During these years, the electrification of the North Island developed at the command of the government. It relied on the land, funding, and organization of the Public Works Department because that is how those in power mapped the process of electrification in New Zealand, starting with Lake Coleridge. As the electric power system grew more extensive, settler engineers and politicians maintained that only a centralized state government could effectively manage the electric power supply. Between 1920 and 1930, New Zealand's total population grew from 1.2 to 1.5 million, and the Māori population grew from 53,000 to 67,000, which resulted in a much larger market for electric power. Throughout the 1920s, demand for electric power grew dramatically, increasing by 25 percent in electric sales per year in residential and industrial spaces.[10] The expansion of the electric grid and other public works reflects the country's renewed national identity and desire for increased cohesion forged by World War I.[11] As public works officials like Julius Vogel had argued since the 1880s, public works infrastructure provided the opportunity and incentives necessary to forge a nation—a settler nation, anyway.

Besides the bureaucratic push for such infrastructures, New Zealand's population, both settler and Māori, also began a significant urban drift during this period.[12] Throughout New Zealand's colonization, settlers increasingly moved to urban centers. The country's export profits increased, primarily due to demand by the British government. The boom-and-bust cycle created by this economic environment encouraged New Zealander laborers, primarily settlers, to move into cities and growing suburbs.[13] Furthermore, the allure of cities like Christchurch and Wellington that now boasted electric light and power spurred urban interest. During the early twentieth century, the heads of the PWA argued that electricity should not remain limited to central urban spaces and should branch into rural regions. However, electric power was mostly confined to the cities outside the agricultural centers. This tended to mean that electric power often did not extend into Māori and immigrant communities.[14]

This pattern of urbanization created some problems. As people moved to fill manufacturing and shipping jobs in cities, rural industries like the dairy and meat markets faltered because of a lack of laborers. The country's leadership and public works administrators believed that providing electricity more readily through rural New Zealand, on both the South and North Islands, could restore balance to the population movement. In a 1927 speech given on hydroelectric developments throughout the dominion, in which he termed New Zealand "the Switzerland of the South," Prime Minister Gordon Coates (1878–1943) argued, "From a national point of view, the most important result which is anticipated in New Zealand from the development of hydroelectric power is its extension to the farming and dairying districts. In common with other countries New Zealand is experiencing an 'urban drift,' and it is expected that the provision of the conveniences and comforts of electric power in the country districts will do something toward checking this drift."[15] Throughout the speech, Coates outlined and explained how electricity could improve the production of New Zealand's primary industries, especially dairy. But, because of overdemand in cities, more power stations needed to be built.

During the Depression years, which in New Zealand many historians understand to have begun prior to the market crash of 1929, the growth of the grid slowed, which left thousands who relied on Public Works expansions without employment.[16] A series of natural disasters and grid failures led politicians and engineers to decide that any expansions required a more substantial regulatory hand and to consider alternatives to hydroelectric systems. Incidents such as the Napier Earthquake (1931), which decimated many of the North Island's existing systems, solidified engineers' insecurities about reliance on hydropower and the ability of regional power boards to handle massive failure. Following the destruction at Napier, electrical engineers and public works officials sought to diversify the means of electric power production and construct additional transmission lines to connect the country's isolated grids. By adapting the existing infrastructure to be more versatile and resilient, New Zealand could maintain its electric system and not rely primarily on one method of power production. In addition to expanding generation methods, the government reinforced the centralization and control of power production using the Electric Power Boards and Supply Authorities Association Act 1930. The act mandated that supply authorities distribute power within the national system by not only forcing them to be licensed by the PWD but also forcing them organizationally and financially into the same corporate body.[17]

By 1939, many power line connection projects were completed, providing 91.7 percent of the country's electricity. Still, electric power was not a universally distributed commodity. Although the power lines, minerals, and water required to power the grid extended onto Māori land, Māori rural areas and towns had not yet been electrified. Despite the uneven distribution of power, the state felt that the electric power system demonstrated a successful long-term investment.

In a speech before Parliament, the minister of public works, Robert Semple (1873–1955), celebrated the year's profits and the relative affordability of electricity compared to wealthier nations like Great Britain, saying that "we can justly claim success for our various electrical undertakings."[18] The PWD planned to celebrate the expansion at the Grand Centennial Exhibition, which would run from November 8, 1939, to May 4, 1940.[19] However, on September 3, 1939, World War II halted New Zealand's plans for development. Parliament quickly passed the Emergency Regulations Act 1939, effectively limiting the production and sale of electricity.

Despite the slowdown in development, the PWD aimed to connect electric power schemes around the country, such as the Mangahao-Waikaremoana and Waitaki Schemes. Alongside the success of many new hydroelectric schemes, however, the harsh realities of war and natural disaster during this period fostered increasing concerns about New Zealand's power security. The stress placed on existing power suppliers resulted in widespread initiatives to encourage New Zealanders to conserve power at home and "Switch off!"[20] As early as 1942, the war began to severely affect materials, labor, and finances for transmission lines and new facilities. The Hydroelectric Branch of PWD encouraged the public to save power, reduced public lighting hours, and restricted the sale of electric appliances.[21] As part of these broader institutional shuffles, on March 16, 1943, the Labour Government restructured the PWD and created the Ministry of Works, which coexisted until the government folded the PWD into the Ministry of Development in 1948.[22]

Much like the Depression, war impressed upon everyone the need for increasingly secure and reliable electric power, which ultimately led to the founding of the State Hydroelectric Department (SHD), initially part of the Electricity Act 1945 and then as an official department within the PWD in 1946.[23] The PWD charged this subdivision with the responsibility to generate power, regulate transmission, sell bulk power, and resolve the country's growing anxieties about energy security.[24] After the Depression and World War II, the SHD expanded the grid and its authority over the country's electric power. Following World War II, which put a considerable strain on the country's resources, workforce, and productivity, many state agencies, including the SHD, worked to evaluate the status and potential of their respective infrastructures. In 1948, Robert Semple, a Labour

Party leader and the head of the MoW, delivered a report to Parliament that highlighted the fragility of New Zealand's grid because of its reliance on other countries for fossil fuel during shortage. Semple emphasized the necessity of making their electric power system more independent: "The shortage of oil fuel is world wide, and there is no reliable information as to the possible duration, or ultimate extent of the shortage. The event, however, does serve to emphasize the importance of keeping our country's vital power supply independent of imported fuels to the greatest possible extent."[25] Although Semple went on to celebrate the potential of hydropower and the foolhardiness of relying on coal for large-scale electric power production, he lamented the inability of the current power supply to keep up with demand. He highlighted a central talking point for those interested in electric power expansion projects: "Electric-power supply is a vital necessity in modern times as the whole economic life of our country depends upon the provision of an adequate supply, and I feel entitled to ask for the active assistance of all members of the community in our efforts to accomplish this task."[26]

Due to new postwar international relations and United Nations participation, New Zealand achieved a new political status by passing the Statute of Westminster Adoption Act 1947 and the New Zealand Constitution Amendment (Request and Consent) Act 1947. The legislation gave the state new powers to control foreign policy, amend the constitution, and structure Parliament. Although some historians place complete sovereignty later, with the passage of the Constitution Act 1986, the 1947 law marked a new degree of independence, which emboldened New Zealand's government to prove its effectiveness, much as electric power systems such as Lake Coleridge became the proving ground for many in the PWD. Furthermore, the government hoped to expand electric power networks to New Zealand's more rural and resource-producing areas by forming the Rural Electricity Reticulation Council. The five-member council represented the industry's generation, transmission, and distribution branches. They organized the levy for the ninety-five electric supply authorities in rural areas. With this new push for rural electricity, the government adopted new tactics for distributing power in isolated places in connection. One of the systems developed was Single Wire Earth Return (SWER). Unlike traditional distribution systems, SWER used one wire to supply single-phase electric power. The distinguishing feature is using the earth, or sometimes water, as the return path to avoid needing a second or neutral wire.[27] The system was successfully implemented in many areas and helped revive electric power usage in the sheep industry, which represented a major part of the country's export economy.[28] During the 1950s, power boards implemented the plan in the Bay of Islands, King Country, Wairere, Banks Peninsula, and Otago.[29]

Despite these expansions, the rapid growth of cities like Auckland required more power than the existing systems could provide. Given recent political shifts and concerns about the continued expandability of hydropower, the MoW began to explore the possibility of using other generation methods. MoW officials and legislators unsuccessfully attempted to incorporate the coal and natural gas industry into electric supply authorities in the Electricity and Gas Co-Ordination Act (1956).[30] In 1958, reflecting New Zealand's energy supply diversification, the SHD became the New Zealand Electricity Department (NZED).

In addition to wrestling with how best to supply the country's growing electric power demands, using New Zealand's resources, between 1950 and 1965 MoW officials sought to connect the country with massive high-voltage lines. This shift would help avoid power cuts and create a less wasteful system that electric supply authorities could centrally manage rather than rely on patchwork clusters of power boards to manage locally.[31] This reorganization drew further attention to the need to consider the limits of hydropower on the North Island, which at least could be partially helped by connecting the North Island grids to the South Island. In 1950, the SHD's chief electrical engineer argued that by 1959 New Zealand would max out its hydroelectric capabilities and began advocating for high-voltage DC links across the Cook Strait to connect the supply of both islands. After extensively debating how to solve this problem, Parliament and the MoW agreed to build the line, but the project did not pass because the new minister of works, Hugh Watt of the Labour Party, believed keeping the grid separate yielded more positive outcomes. However, Labour's control of the government was short-lived; in 1961 the National Party resumed the Cook Strait Project, and it was completed in 1965, officially connecting both islands. The resumption of this project highlights the preference of the National Party—in charge for much of Huntly Power Station's construction—and much of the rest of the country to think about electric power demand as something to be expanded to serve the country's consumer and production expectations. Furthermore, it highlights the consistent portrayal of the North Island's hydroelectric shortcomings. The South Island's burgeoning industries and hydroelectric potential left officials and many people in the North Island searching for an answer to the power problem.

## A SOURCE OF POWER FOR THE NORTH ISLAND

In 1867, a century before the construction of the Huntly Power Station, Ferdinand von Hochstetter (1829–1884), surveyed New Zealand's resources and described the energy potential of the area around the North Island's Waikato River, home of the Waikato Tainui Iwi, the largest group of Māori iwi in the region, as a treasure to be exploited by the descendants of settlers. From the earliest days of European invasion forward, the colonial and national governments understood the Waikato region as a resource-rich frontier waiting to be tapped. For example, Hochstetter said of the river and surrounding region, "At any rate, there lies a considerable store of fuel, which will be raised soon as European settlements have commenced to extend over the beautiful lands on the lower Waikato, and steamers to navigate the river. It is a rich treasure reserved for generations to come."[32] Not only was the space to be exploited; it was also to be exploited for settler designs for New Zealand. The area sat in the King Country, Te Rohe Pōtae, or the Western Uplands, a loosely defined region established during the New Zealand Wars as colonial troops sought to quell the Kīngitanga (Māori King) Movement. Hochstetter's comment signaled the potential value of the resources buried within the ground in Waikato, including coal. The survey typified the language and opinions of European colonizers in New Zealand who saw the islands as a resource frontier, perhaps the last one to be exploited.

The first mention of a power station in Huntly came in 1918, when the town boasted a population around 1,535, including about 150 Māori. The idea for a significant plant at Huntly to power the North Island appeared in a report Parry submitted to William Fraser, the minister of public works: "The erection of a power-house in the Huntly district has been suggested as a suitable source of electric power for the Auckland district, the plant being designed to burn the slack coal produced in the mining the more marketable coals."[33] Later in the report, Parry weighed the benefits and objections to such a station.

Namely, he worried that the poor quality of some of the surrounding coal mines could prove bothersome. However, he maintained that the cheapness would likely pay off in the end, particularly for Auckland's growing power demand. Despite this mention, hydropower remained the favored source of electricity, mainly because of the control the MoW maintained over the water systems of New Zealand.

While the Lake Coleridge discussions intensified, electric power generation efforts on the North Island took a different approach and expanded electrification. The Waihi Gold Mining Company spearheaded this effort and became one of the country's top industrial electricity users.[34] Due to the passage of the Water-Power Act 1903, at the time they needed approval from PWD, which they received via the Public Works Amendment Act 1908—constructing the Horahora Power Station (1913) on the Waikato River and generating power. Beginning with Horahora and following with the stations built by the power boards in the region, Huntly started to receive electric power for lighting and certainly heard about electrification projects through New Zealand's press. However, the electric power supply on the North Island remained largely fragmented until 1934 when the MoW, under John Bitchener (1864–1952), connected major power lines between the Arapuni and Stratford power stations. By the decade's end, plans for further electrification intensified as successive PWD (and eventually MoW) statements reported growth in demand and profit on the North Island. Conversations about Huntly as a possible location for a power station continued well into the 1940s and 1950s but were largely abstract until the 1960s.[35]

The first concrete proposal for a power station at Huntly appeared in 1969, in a report submitted by the Energy Planning Committee of NZED. The bid mentioned the potential of building a large 1,000-megawatt power station, comparable to but slightly smaller than contemporaneous projects like the Grand Coulee Dam in the United States or the Volga Dam in Russia. Such a station would deliver more electric power than all four dams on the Waikato combined. The authors first advocated for a nuclear project, but quickly shifted to thermal as New Zealand politicians and activists increasingly pushed against nuclear power plants.[36] Eventually, NZED settled on a thermal power station so that they might rely more heavily on domestic power sources. The government considered a few locations for the station, but the mayor, T. Gavin, and the Huntly City Council effectively promoted their town. As a collapsing coal-mining town of approximately 5,000, they sold NZED on the availability of coal and natural gas that would grow the population, regenerate the local economy, and bring new people to the region. And for many people in Huntly, including Tainui and settlers, the promise of a significant construction effort and power station seemed a promising source of employment and growth. A new mining workforce and massive influx of construction laborers would not only increase the population but also expand the existing demand for goods and services that could rejuvenate the town.

Huntly, like so many other towns, had developed around extractive industry. Like Reefton, it mirrors the British colonial vision of New Zealand by existing to exploit its resources and make the islands' settler population increasingly self-sufficient. They did this, of course, through the forceful removal of Māori and the theft of their land. However, unlike in Reefton, in Huntly the Māori had settled permanently and more densely. Prior to European invasion, Huntly, or Rāhui Pōkeka, and the surrounding region now known as Waikato were the home of the Tainui, a confederation of Māori iwi who trace their lineage back centuries.[37] The Waikato River, which runs through the region, is regarded as the living ancestor of the Tainui; sometimes they refer to themselves as the Waikato Tainui Iwi. According to Māori and European sources, the river provided ample food, agricultural opportunities, transportation, and communication networks. Following the Treaty of Waitangi (1840) and manipulative land acquisition procedures developed by settlers, the Tainui refused to submit to the Crown and clashed with the British during the Kīngitanga Movement, which they and other iwi organized to halt the settler and government alienation of Māori land. The Tainui and Auckland-based governments' conflicts resulted in some of the most brutal battles during the New Zealand Wars (1845–1872). The British forced many Tainui involved to leave the region or give up their lands in Waikato through violent and exploitative land confiscation through the Suppression of Rebellion Act 1863 and New Zealand Settlements Act 1863. The government returned some lands in the 1920s as Land Grants and in the 1990s following several Waitangi Tribunal hearings.[38]

The Tainui involvement in the Kīngitanga Movement and New Zealand Wars created a seat of Māori political and cultural power in Waikato. For example, a few kilometers outside of Huntly sits the Waahi Pā. *Pā* can often mean "hill fortification," but in this case, it refers to the more generally used meaning, "settlement." The pā is on the banks of the Waikato and serves as the meeting place for the Ngaati Kuiaarangi, Ngaati Mahuta, Ngaati Tai, and Ngaati Whaawhaakia hapū (sub-tribes) of the Tainui. The space functions as a marae. Usually the shortened form of marae ātea, marae means a place where culture can be celebrated; often, it colloquially refers to the area near or around the wharenui, a meeting house. The Waahi Marae functioned as the seat of power for many Kīngitanga leaders and Tainui Māori who worked to advance Māori rights from the late nineteenth century.

During those conflicts, the settler community maintained a small settlement on the Waikato River south of Auckland that eventually became known as Huntly. The station was a military outpost during the Waikato Wars (1863–1872). In the 1870s, with the designation of a postmaster and speculations about deep coal seams in Waikato, settlers began to refer to the growing settlement as "Huntly Lodge."[39] The officer responsible for the designation, Captain Anthony Ralph, received land in the area for his military service, and founded the Taupiri Coal Mining Company.[40] The coal mine flourished. In an 1877 parliamentary memorandum on coal, Allan V. MacDonald celebrated the quality of the coal from this mine because it "seldom has clinker," or non-burning slag.[41] Miners flocked to the region from the South Island and Scotland, and many Māori displaced by the Waikato Wars and land theft also worked the mines.[42]

Coal affected the lives of all who lived in Huntly and the surrounding area. The industry flourished in the 1870s and 1880s under Julius Vogel's public works initiatives. By the late 1890s, coal works had expanded across the country and become one of its primary energy sources, a statistic that would not recede until the postwar years. Coal powered the trains that Vogel hoped would connect the country and increase the potential flow of steamships into New Zealand. Domestic industries like freezing, meatpacking, lighting, and heating all relied heavily on coal. For the remainder of the nineteenth century, the mine at Huntly remained profitable and fed New Zealand's growing locomotive and steamship industries. By 1898, it produced around a hundred tons of coal a day.[43] Coal mining expanded in the town well into the early twentieth century, despite the mining strikes in 1913 and the tragic explosion on September 12, 1914, that resulted in the death of forty-three miners.[44] As the city grew, water mitigation infrastructure expanded, and a small farming community formed in and around Huntly. A few other small industries like brickmaking flourished, but Huntly remained a mining town well into the mid-twentieth century, with opencast (open-pit) mining beginning on both the east and west sides of the Waikato River in 1945.

From the 1920s onward, even though it remained in demand, coal purchases shrank, especially by domestic consumers, in no small part due to the expansion of the electric power network. Amid significant revisions to the electric grid, mining in Huntly and the surrounding Waikato mines faltered. People and industries preferred electric power, and fewer mines and miners were required due to changes in mining technology, such as mechanized draglines, power shovels, and loading equipment. Furthermore, the increased usage of diesel and petrol replaced sources of coal consumption. Despite the decreased usage of coal, it remained a key extractive industry in New Zealand, because the colonial and later the Dominion government made energy resources a central piece of their governing logic. One of the notable features of the coal mines in New Zealand is that, much like hydropower at Lake Coleridge, the government, even before Dominion, insisted that state control of the resource was necessary for maintaining a prosperous state. Richard Seddon, prime minister of New Zealand from 1893 to 1906, created the State Coal Mines to compete with other coal monopolies. As this happened, the coal industry briefly recovered, and mine works in the Grey River Valley and the Waikato region flourished and continued to do so until after World War II. By the late 1940s, the Labour Government had nationalized the mines, including those in Huntly, the future site of New Zealand's largest thermal power station to date. That changed marked a new era of coal mining using new opencast mining methods and an expansion in coal production that peaked in 1960 and undoubtedly stoked the appeal of Huntly as a site for electric power production.

## THE FIGHT AROUND HUNTLY

From the beginning of the planning process, the treatment of the environment defined how people thought about, advocated for, and fought against the Huntly Power Station. In 1971, NZED, much like the Public Works Department had done before the construction of Lake Coleridge, surveyed the country's resources and consulted interest groups. However, unlike with the Lake Coleridge scheme, the government mandated that NZED produce an environmental statement and continue to comment on it throughout the process. NZED had to consider the environment because of the growing national interest in protecting resources, bound both to sincere conservation efforts and decades-long resource control practices disguised as such and advanced through the legislature. Following a series of laws passed in the early twentieth century, such as the Scenery Preservation Act 1903, the Public Reserves, Domains, and National Parks Act 1928, and the Soil and Rivers Control Act 1941, New Zealand, in step with the global environmental movement, formed new bodies to investigate and control government and corporate actions that could prove detrimental to the health of people and ecosystems.[45]

During the 1960s and 1970s, New Zealand politicians and activists participated in many of the emerging environmental movements. In response to major MoW efforts—namely, the Manapōuri Power Station—Parliament implemented the Nature Conservation Council Act 1962. The act created the Nature Conservation Council, whose functions included a scientific and technical review of public works projects and assessing their impact. Additional advisory bodies, such as the Physical Environment Committee (1970) and Environmental Council (1970), were created during the 1970s to consider public works and draft environmental standards. However, these entities possessed no statutory power until 1972, when Prime Minister Keith Holyoake created the cabinet position of minister for the environment and the Commission for the Environment. Despite the title of minister, the position and committee lacked clear guidelines on the agencies with whom they could interact or influence, such as NZED. As Faeeza Ballim recently argued in *Apartheid's Leviathan*, a study of South Africa's electric infrastructure, vast arrays of ambivalent technological and regulatory bureaucracies wield an extraordinary amount of influence.[46] As other historians have noted, while such bodies represent essential milestones, they did not maintain meaningful authority or clear jurisdiction when it came to advocating for environmental protection in large projects—especially, in the case of New Zealand, during the construction of the Huntly Power Station.[47]

Despite its statutory limitations, one of the significant products of the Commission for the Environment was the Environmental Protection and Enhancement Procedures (EPEP) in 1974. Following the concerns raised during the construction of the Manapōuri Hydroelectric Scheme and growing concerns about environmental harm, drafters intended the EPEP to give decision-makers insight into major government projects to determine whether they might cause significant environmental issues. However, the statements submitted by NZED were not yet bound to this procedure. It was a new endeavor, which NZED admitted at the beginning of their 1972 Environmental Impact Report: "Since there has been no previous requirement for an environmental impact statement, this first statement includes a summary of the work which has been done to date, which is pertinent to the environmental impact of the proposal and an indication of the area in which further investigations are required."[48] J. A. Walding, the minister for the environment, insisted in an early commentary on the impact report that the environmental issue "is, I believe, an issue crucial to the Government's relations with the environmental movement."[49] The NZED Environmental Impact Statement of 1972 noted that "the majority of questions asked are concerned with air and water quality."[50] Following approval from the treasury in May 1973, the final approval to build this station came on August 13, 1973, making cursory references to paying attention, when the builders could begin, to the suggestions made by the Commission for the Environment.[51] The cabinet meeting minutes show that NZED planned to "arrange studies" of the station's impact on the Waikato River's water level, its temperature, and the region's wildlife. The environmental effects were to be assessed as the station was being built. Notably missing from these impacts are the social and economic effects the station and its construction might have on the people there.

Engineers and public works officials found the existing language in the proposal about environmental protections ill informed and superficial. For example, B. G. Powell of the New Zealand Institution of Engineers called the content in one of the impact statements "seemingly superficial information."[52] Beyond Powell's critique of the completeness of the information provided by NZED and the MoW, he and others noted the need for more availability to the public. The Environmental Defence Society (EDS), a not-for-profit organization, was founded in 1971 to push against what many activists deemed as problematic environmental policy. Its members, whom MoW Committee for the Environment Chair Robert G. Norman called the "greens" in his memoir, fought for access to MoW records and the

FIGURE 4.2. Megawattasaurus. Sid Scales, "Help—a Monster!" *Otago Daily Times*, September 19, 1975. Alexander Turnbull National Library, https://natlib.govt.nz/records/31983031.

Environmental Impact Report from 1972 until the station was completed.[53] The government couched their delays in making information accessible by citing the public's inability to comprehend the technical specifications of an advanced thermal power station. For example, in 1973, the deputy prime minister, Hugh Watt, hinted to the *New Zealand Herald* that the government was working to develop the reports in a way the public might understand.[54] On June 6, 1973, continued public pressure forced NZED to release the MoW records containing the new power station's environmental impacts and implications, which opponents saw as the gateway to halting the construction.

These regulatory developments illustrate how, even amid vast shifts in thinking about the environment, power production remained the most central concern for MoW officials. Regulatory bureaucracy was an annoying hurdle for the MoW and the electricity industry; the latter appears in figure 4.2 as a "megawattasaurus" overreacting to oversight by the Commission for the Environment. Ill-defined policies and lack of public transparency represent the MoW's intentional decision to slow resistance to significant construction efforts. The memos and notes of the MoW's Committee for the Environment chair, Robert Norman, frequently mention that while they remained prepared to consider long-term environmental effects, the looming power shortages and potentially high cost of foreign energy resources remained the MWD's main priorities. For people like Norman, the cost of the delays created by the environmental review process did not impact those pushing against development, instead burdening the MoW. Even twenty years later, when Norman wrote his memoir, he maintained that the risk of not having enough electric power outweighed the potential dangers of environmental damage that might be discovered; human costs remain absent in his memoirs, too.[55]

## TAKING LAND TO GENERATE POWER

In response to the lack of concern and overt exclusiveness with which NZED and MoW handled the station, Sir Robert Te Kotahi Mahuta (1939–2001), Tainui Rangatira (or leader with hereditary ties to the iwi), spoke against the flawed reports and assessments. In addition to acting as a leader for the Tainui, Mahuta had been appointed in 1973 the director of the Centre for Māori Studies at Waikato University and had studied anthropology at the University of Auckland and Oxford. He thus was well situated to navigate settler discourse on infrastructure.[56] Over the years, Mahuta published and presented numerous papers about the inequity and harm wrought by the machinations of NZED and MoW to build Huntly. During the episodic release of the environmental impact statements mentioned earlier, Mahuta and other Tainui frequently complained about the lack of reporting on the station's social impact and the absence of transparency about the MoW's findings. Mahuta asked, "What kind of society are we really trying to create in this country? Another America, another Japan? Here in Huntly, the questions as far as the Māori people are concerned are—Progress, at what price? . . . Energy development can no longer rest on narrow technical expertise but must take account of the social and environmental implications."[57]

Settlers portray Māori communities, particularly during this period, as fundamentally opposed to development or technology due to their advocacy for environmental reform and new land-reclamation

963

**JASMaD research**

Stephen Jelicich · John Austin · Graham Smith · Ivan Mercep · Rodney Davies · John Sutherland

P.O. Box 6648 Auckland 1 New Zealand
2 Whitaker Place.
Telephone 30497 Code JASMAD

RECEIVED
22 MAR 1973
by

2000/RJPD/SCW

19 March 1973

The Commissioner of the Environment
Environment Commission
Private Bag
WELLINGTON

Attention: Mr P.J. Brooks

Dear Mr Brooks

You will recall that I wrote to you at the beginning of November introducing you to our organisation and expressing our interest in environmental assessment studies. At that time we suggested that we should meet the next time you were in Auckland, but this we have not yet managed to do.

The purpose in writing now is with reference to the proposed Thermal Power Station at Huntly. I enclose a copy of a newspaper article cut from the Auckland Star of the 2 March 1973, with which you are no doubt familiar. You will see that both Mr Dart and Mr Mahuta call for environmental impact studies. Mr Dart suggesting this should be done by the University of the Waikato where Mr Mahuta is Director of the Maori Studies and Research Centre. Again, in order to be credible, neither the University or the Electricity Department are in a position to provide an objective and unbiased assessment.

The question of the use of Maori land for development works is a contentious one. Often it appears to be under-utilised, but to the Maori it symbolises more than just land use potential - his whole mana or status in society is affected.

Obviously my purpose in writing is not just to be chatty. I have travelled to Huntly and inspected the site. The conditions for the School are as bad as described and this situation does appear to call for a study of both the social and physical aspects of the development. We believe that we could assist your Commission in this matter.

Yours faithfully
JASMaD RESEARCH

Rodney Davies

R.J.P. Davies

Enclosure

FIGURE 4.3. Letter containing Davies's quotation. Rodney Davies to P. J. Brooks, March 19, 1973, "JASMaD Research Memorandum Concerning the Environmental Assessment," Huntly Power Station Reports 1972–1973, Ministry for the Environment, R1281876 Box 69, Archives New Zealand Te Rua Mahara o te Kāwanatanga, Wellington.

policies. This connects with a well-practiced pattern of assigned placelessness deployed by settler states worldwide that aim to delegitimize Indigenous claims and authority.[58] These assertions mischaracterize Māori positions and ignore more than a century of Māori engagement with the material economy of the settlers.[59] This is especially true for the Waikato Tainui involved in the Kīngitanga Movement, founded to create a pathway through which Māori might retain agency over their land, water, and other resources. Although many Waikato Tainui would later object to the construction of the plant, at first many hoped, much as the settler population did, that the Huntly Power Station might revitalize the community. Few reports of early exchanges among officials make mention of the Māori, but some external parties did write to the planning committees. For instance, on March 19, 1973, Rodney Davies, an architect with JASMaD, a firm that worked with many Māori communities on the North Island to restore buildings and incorporate Polynesian architecture into Auckland, wrote to P. J. Brooks, the commissioner for the environment: "The question of the use of Maori land for development works is a contentious one. Often it appears to be under-utilised, but to the Maori it symbolizes more than just land use potential—his whole mana or status in society is affected" (figure 4.3).[60] Whether Davies's motivations were profit driven or genuinely concerned with the threat the power station posed to the Tainui is unclear. What is clear is that sociocultural considerations were made early in the planning stages by Māori and settlers.

Despite these considerations, however, the NZED ultimately chose a location near the Waahi Pā Marae for the power station (figure 4.4). This marae functioned as the center of the Kīngitanga Movement and, at that time, was home to Queen Atairangikaahu (1931–2006). Despite the Waikato Wars, land confiscation, and colonial policy-making, the marae allowed the Tainui to maintain a place of refuge, spiritual sustenance, political stability, and social cohesion. The Waikato River, which flows right next to the marae complex, provided key economic resources and food and, noted Mahuta, the "physical embodiment of tribal mana and spiritual prestige."[61] How, he asked, could such a place not receive public funding or consideration when building something as significant as a 1,000-megawatt power plant? "One wonders whether the general community would support the building of a power station next to the Governor-General's residence or the home of similar pakeha [*sic*] dignitaries."[62]

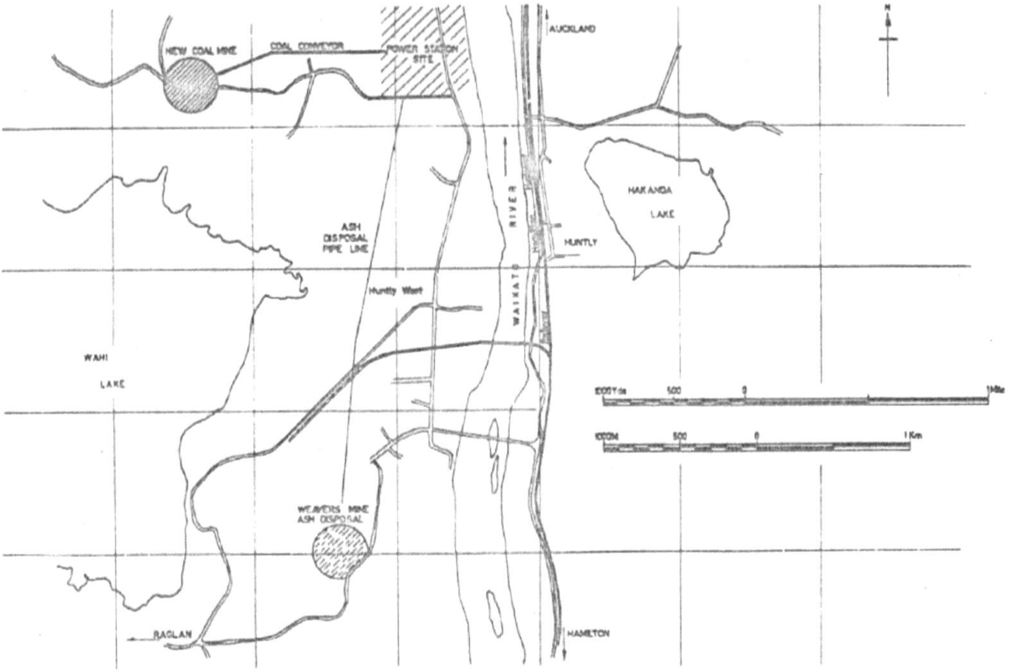


FIGURE 4.4. S. J. Morrison, Huntly Power Station: Proposed Location, New Zealand Electricity Department, Memorandum for Cabinet Works Committee, 1973, Appendix, Archives New Zealand, R1281876

However, what mattered far more than the plant's proximity was the marae's history and cultural resonance. The history of the Māori people is replete with examples of settler communities and officials dismissing and often willfully remaining ignorant of the importance of marae to communities. The dismissal of Tainui culture and beliefs led to the dismissal of their protests over water rights and environmental degradation. Perhaps the most potent example of this emerged in the Environmental Impact Report's designation of the Waahi Pā Marae as a structure that was not a "public amenity" under the Electricity Act 1968.[63] Building on earlier MoW practices, the act defined an *amenity* as "those qualities and conditions in a neighborhood which contributed to the pleasantness, harmony, and coherence of the environment and its better enjoyment for any permitted use." NZED argued that the Waahi Pā Marae was not freely open because cultural protocols existed on when and how the space was to be used. This disqualified the marae and community from receiving an entitlement grant of 1 percent of the profits from the power station. Tainui raised complaints about this interpretation. Mahuta called the designation "predictably subjective and discriminatory."[64] The Waahi O Marae Committee, on behalf of the many hapū members in the region, argued that NZED's designation failed to understand that a marae is a public place and a public facility, stating, "We maintain that it is a public amenity . . . the particular role of the marae as a democratic forum of discussion and as a ritually appropriate place for the whole of the Huntly community, not only the Māori community."[65]

After addressing the marae's claims to the prime minister, Mahuta received a memo from P. W. Blakey, the general manager of NZED, addressed to the *Auckland Star*, pointing out that many of Mahuta's claims did not directly link to NZED's activities and chastising him for not addressing the appropriate offices, saying, "Furthermore I am informed that no complaints from people affected by the present work have been directed to the District Commissioner of Works in those area [*sic*] the site is. It is unfortunate that the article, as written, tends to give a misleading impression of the extent to which the New Zealand Electricity Department is involved in activities at present taking place in the vicinity of Waahi pa." Blakey also pointed out that two projects were taking place in Huntly on the Waikato: power station surveys and construction of a stopbank to reduce erosion and safeguard agriculture. Mahuta responded by reminding Blakey that regardless of the projects' governing body, both were related because they disrupted Tainui land, culture, resources, and building. He also argued that "the unhappy experience of the Waikato people in their dealings with bureaucracy has led to the situation where they choose to present their grievance directly to the Government of the day. Certainly none of the local residents would even be aware of the District Commissioner of Works, his powers, or the way in which they could communicate their concern to him."[66] For Mahuta and the Tainui, the intricate bureaucracy of supposed social and environmental protections managed by the MoW had enacted the widespread disenfranchisement of Māori autonomy.

Another demonstration of the MoW's lack of consideration and disregard for the Tainui during the construction of the Huntly Power Station is the Rakaumanga School, or Te Wharekura o Rakaumanga. As part of a broader education and assimilation program advanced by the Native Department and later the Department of Education during the late nineteenth century, the Rakaumanga School was one of hundreds of schools designed to provide primary education, especially in English, to Tainui children. Though a colonial institution, the school became an essential community space through which many Tainui leaders fought to retain their identity and culture, especially when the New Zealand government hoped to dissolve the school and integrate Māori children with those of settlers. In fact, during the 1970s after integration, when Huntly's construction began, Tainui fought to retain the instruction of Te Reo Māori (the officially recognized and alphabetized Māori language) and Tainui history in the school.

Rakaumanga School sat near Huntly's building site on Te Ohaki Road. As parents of students lamented, the construction created noise pollution and dust, and potentially would force longer commutes to other schools that might be overcrowded, receive less funding, and ignore Tainui culture in their curriculum. The authors of NZED's 1972 Environmental Impact Report mentioned but dismissed the school's significance: "It would seem that this school is regarded locally as a 'Maori' school and the wish of the elders is to keep it so. In fact, the school does not have any special designation as there are no longer any Māori schools, other than some private schools, and the Education Board have, from some time, been planning to relocate the school further to the south nearer a new housing area."[67] Many local politicians recognized that moving a school for a public works project conjured violent memories of colonial rule. For NZED, it was another logistical issue to overcome. For others in the government, including Prime Minister Norman Kirk (1923–1974), it garnered national attention because of the four Māori seats that the Labour Party held. Still, NZED proceeded with construction.

The disregard for community concerns illustrates how engrained the decision to proceed with construction was with settler ideas about erasing and subsuming Māori through settlers' interpretation of the law. With few exceptions, MoW and NZED officials viewed Māori issues as little more than inconveniences. They remained unconvinced by the overt similarities between their actions and the land confiscation policies of the late nineteenth century that Tainui and other activists had indicated. From that period into the early twentieth century, the government passed many acts like the Fisheries Conservation Act 1884 and the Land Drainage Act 1893. The former restricted what people could fish for and when, empowering the government to "make, alter, or revoke" needed regulations. The latter enabled the MoW to drain lands for agricultural purposes, including any "Native Lands."[68] While these policies indicate conservation efforts, they also signal a settler-colonial agenda that saw conservation as a mechanism for ensuring economic vitality and control over the land, which fit many of the British colonial and settler state doctrines of development during the twentieth century.[69] Beyond preserving the environment and ensuring agricultural productivity, the MoW needed water control to facilitate their expanding hydroelectric networks. Increasingly in the 1900s, the MoW implemented legislation that granted them the ability to tightly control water resources, such as the Soil Conservation and Rivers Control Act 1941.

More than eroding the potential for economic enrichment for both the settler and Waikato Tainui communities, the site selection process for the Huntly station recatalyzed century-old land acquisition disputes. NZED needed to obtain land for both the power station and the town it was planning for its builders, including five hundred new homes and associated facilities. In November 1973, a few months after the approval, Norman McLeod, commissioner of works, reported that only one block of land, referred to as Kimihia Farm, met the requirements, room for houses and proximity to the power station. Some in the department quarreled about the viability of this space, but McLeod and the leadership stuck to it. The land owner, Rosser, did not want to sell the land to the Crown, preferring to sell it to a higher-paying private developer. McLeod advised the relevant departments that the government could rely on the Public Works Act 1928 and the Electricity Act 1968 to make land sales compulsory. It is worth noting here that this directly contradicted NZED's promise to the community, Māori and settlers alike, that they would not be taking land.[70]

The Public Works Act had been used many times before to acquire land, and often the compensation did not reflect the market. MoW officials used section 276 of the act, which stated, "Land required for water-power or irrigation works or purposed may be taken as for public work."[71] Furthermore, this type of control over resources is etched into many laws allowing the government to construct public works. For example, the Town and Country Planning Act 1953, despite codifying the mechanisms through which the government could build, improve, or tear down existing spaces in New Zealand, does not mention the word *marae* or include language that might safeguard them. NZED gained the authority to build the power station through three critical pieces of legislation: the Water and Soil Conservation Act 1967, the Electricity Act 1968, and the Public Works Act 1981. The use of these legislative vehicles highlights how NZED and the MoW overcame the hurdles introduced by protesters and Māori, objections from within the government, and multiplying environmental protection protocols to fulfill the government's desire to overcome international anxieties about energy independence in the wake of the oil crisis and prove New Zealand's political resilience.

Although Tainui concerns ranged from environmental degradation to diminishing agricultural prospects, civic infrastructure, and civil rights, they also highlighted the issues of transparency and challenged the environmental data about the Huntly station's effects on water temperature and pollutants. In the plans for the power station, water drawn from the river enveloped the boilers, heated to steam to power the turbines, condensed in the cooling towers, and returned to the river. One of the central concerns that protesters and environmental groups raised was the temperature of this returned water and its potential impact on the river. As Robert Norman and other MoW and NZED officials lamented during the planning process, NZED had to gain approval to use water from the Waikato River and reinject it.[72] The Water and Soil Conservation Act 1967 required NZED to obtain permission from the Waikato Valley Authority, a regional government organization managing the river and its subsidiary waterways, flooding, agricultural water usage, and sediment usage. The 1967 act declared that public works had to "make better provision for the conservation, allocation, use, and quality of natural water; the promotion of soil conservation; the prevention of flood risks; the regulation of the multiple uses of national water in a way that took adequate account . . . of the needs of primary and secondary industry and to ensure the continuity of water supplies of local authorities, fisheries, wildlife habitats, and all recreational uses of natural water."[73] In its 1973 proposal to the MoW, NZED insisted that Huntly was of such national importance that the authority was obliged to ensure the success and security of New Zealand's energy future. The Waikato River Valley Authority's chief engineer rejected this and forced NZED to pursue water rights permits.

Throughout the project, the rights to use and discharge water changed hands numerous times. When it came to the plant's construction, confusion existed about who possessed these rights. Although the initial applications came from NZED, the MoW also possessed the rights at some points during the construction process. This is a potent example of how the newly formed environmental protection policies and guidelines did not seem to apply to the Huntly Power Project, and how the desire for power and energy security dwarfed other concerns. Ultimately, the Waikato River Valley Authority determined that NZED needed two permits to adhere to these standards: one to extract water for generation and one to discharge it for cooling purposes. These permits required appeals to the minister of works for permission. Much to the chagrin of NZED, advocates for Huntly within the Ministry of Works, and the construction firms, this required that they perform a biological survey of the river to determine the impact and maintain the river's water levels.

These applications for NZED's acquisition of the right to proceed with their planned usage of the Waikato, numbered 220 and 221, respectively, immediately led to appeals by Tainui and EDS. I will focus on the appeal made by Tainui, done under section 25 of the Water and Soil Conservation Act 1967 and sections 4 and 5 of the Town and Country Planning Act 1953, which grant the appeals boards, comprising government officials, the ability to "vary, cancel, or reverse" a public works scheme.[74] Tainui, through Robert Mahuta, appealed because they argued they were harmed by the granting of water rights to the power plant. Mahuta argued that the Waikato Tainui were charged by custom, tradition, and usage with four roles:

> Charged with the Control of the Right to Use the Waikato River
> Charged with the responsibility to protect the rights of our people to traditional foods such as white-bait, mullet, kaeo, ell, and carp, granted under section 2 of the Treaty of Waitangi.
> Charged with the care and well-being of the Waikato River, which we regard as our ancestor.
> Charged with the protection, preservation and improvement of the community life of our people which in turn centres around the Waikato River.[75]

He also argued that the allocation of water rights most affected the iwi because they lived beside the river and would most feel the disturbance. The Waikato was and remained firmly understood by Tainui as tupuna (ancestor), and the biological consequences of disrupting and removing water extended beyond the life of the river's inhabitants. The Waikato, according to Mahuta, was responsible for maintaining the river's mauri (lifeforce), which comprised the water, people, flora, and fauna of the region and the nonmaterial things that needed to be protected. Mahuta's appeals show that such considerations had not been made during the planning process. That information about the environmental impact had again not been presented to the Tainui, or indeed the public more broadly, which connected to the appeal made by the EDS. The Town and Country Planning Appeals Board overruled the majority of Mahuta's claims but suggested that he might appeal again for increased transparency. And he did.

In 1975, NZED applied for water rights again to discharge water that was used to clean the plant's boilers (rights 469 and 479). These rights were approved by the MoW and then subsequently appealed by the Tainui, as well as the Waikato River Valley Authority, Auckland Regional Authority, and EDS.[76] These appeals, in step with the Te Rōpū Matakite o Aotearoa ("Those with Forsight" or Māori Land March of 1975) and the wider Māori Cultural Renaissance, generated unprecedented public attention and played a significant role in the Waitangi Tribunal's formation.[77] The Tainui appeal encapsulates much of what comes out of the other appeals. This time Mahuta acted on behalf of the Tainui Māori Trust Board, one of the many trust organizations created in concert between Māori and the government to address land confiscation claims by granting compensation through a trust fund. The Tainui Māori Trust Board appeal claimed that their iwi's culture and rights were threatened by granting these water rights to NZED. Mahuta made many of the same arguments about transparency and consideration that he had advanced in 1973 but also insisted that the National Water and Soil Conservation Authority had no right to grant NZED the rights. Doing so, he maintained, had violated one of the critical principles of the Treaty of Waitangi: Tino Rangatiratanga, which means "sovereignty" or "self-determination," depending on the interpretation of the treaty. Many Māori, including the Waikato Tainui, understood the term as the responsibility and capacity to make decisions around water and land. Mahuta and the Tainui Māori Trust Board insisted that such a decision within the law resided with the Tainui, but these appeals also failed. The ambiguities of the environmental protection laws and the absence of Māori rights in public works rendered enforcement and protection moot while also granting power to the government departments to decide which laws to abide by.

The appeals process rhetoric highlights how NZED and the Town and Country Appeals Board misrecognized Māori rights and ties to the environment, privileging subjugation and development. Initially, the Town and Country Planning Committee that heard Mahuta's appeal dismissed it because they deemed it irrelevant to the inquiry. They nodded to the cultural impact the power station might have on the Tainui. They could do this because provisions for Māori values and rights do not exist in the Water and Soil Conservation Act. Apart from failing to recognize these things, this legislation forces the Tainui to ask for recognition, a process that reinforces continued misrecognition and subjugation. Indeed, even as protections and revisions have increased since the founding of the Waitangi Tribunal, and more pointed cases dealing with water and Māori communities have been advanced, like the Kaituna River Document or the Report of the Waitangi Tribunal on the Manukau Claim (the Manukau Report), many Tainui still experience disrespect and injury at the hands of legislation and policy in the region.

## THE SUPREMACY OF ELECTRIC POWER GENERATION

Huntly is located on land stolen through settler-colonial violence, legislation, and understandings of energy. The existence of the power station depended on that. And the repeated ignorance of Māori concerns in the area is what so closely connects this power station to colonialism.[78] NZED's maneuvers in the 1970s do not represent new thinking or an abrupt policy shift. They are instead the product of decades of policy-making that disenfranchised Indigenous autonomy and granted state departments, like the MoW and NZED, more power. Although the history of the decision to build the Huntly Power Station features valuable perspectives on New Zealand's electrification, politics, and environmental debates in the mid-twentieth century, the power station should also be contextualized within the country's longer electrification story, because the station's development and path to construction—despite it relying on a different source of energy, being built under a new Ministry of Works, and the power of new environmental protections in New Zealand—heavily relied on settler ambitions, many of which mirrored the plans and construction of the power stations at Reefton, the Phoenix Mine, and Lake Coleridge.

From its settler-founding, Huntly functioned as an "energy" town in service to the expansion of colonial interests on the island and eventually to the aims of the dominion government.[79] The 425 kilometers of the Waikato River showed potential to serve the Crown's interests through the extensive resource holdings in the area, including access to and transport of mineral resources. Throughout the late nineteenth century, settler culture established Waikato—which became known as King Country, because of the Kīngitanga Movement—as demographically and geographically Indigenous, a place still left to conquer and assimilate. Although the physical violence that defined the space receded in the twentieth century, attempts to tame this "last frontier" and the people in it remained a key feature of public works initiatives.[80] The British portrayal of the area and resources as simply the logical next step of settlement ignores and conceals cruel colonial practices. The control and sale of resources are not simple acts of extraction but a vicious fragmentation of the heritage and spiritual markers of the Tainui. Following the land confiscation acts, the government passed the Railways Construction and Land Act 1881 and the 1883 amendment to the Counties Act, which permitted the ongoing alteration of land and water for settler and colonial purposes. The Public Works Act Amendment Act 1889 and the Water-Power Act 1903 ensured that the nation's water and energy potential remained firmly wrested from the Tainui, especially as conversations turned toward electrification.

As has been well covered in other scholarship, government officials recognized that the connectivity and dependence created by electric power systems afforded the managing parties an extraordinary amount of influence.[81] Well into the mid-twentieth century, settler-politicians and journalists echoed and reinforced colonial electrification paradigms by insisting that the centralization of electric power would yield a more productive New Zealand. In 1940, one journalist celebrated the country's hydroelectric primacy in a commentary on the country's electric history: "The Great Development—the completion of the main State Hydro Stations. . . . Probably nowhere else in the world is the purely rural reticulation so fully developed or better organised than in New Zealand."[82] In 1947 the Hydroelectric Department produced a film through the National Film Unit with dozens of shots celebrating the country's hydroelectric sources, especially the Waikato, with several references to the Arapuni Dam. It focuses on two men who controlled the flow of water through the "flick of a switch" from Hamilton, then showcases the hustle and bustle of Wellington and Auckland and dairies across the North Island, to celebrate the businesses and economic productivity sustained by electric power, which could only be sustained through centralized authority ordering blackouts, shutoffs, and redirections of power supply. The video features Duke Ellington's "I'm Beginning to See the Light," as the narrator explains New Zealand's dependence on electric power and exclaims that "the country needs more electric power and it needs it now . . . the Waikato River will give the people boundless electric energy."[83]

When the state took interest in providing power to Māori communities, it hinged upon the ability of electric power to create opportunities for productivity. For instance, in 1949, as part of the Māori Land Development schemes on the North Island, the Māori Affairs Department made provisions for the ex-servicemen of the Tūwharetoa iwi to receive electric power provisions for the sawmills and dairy industries in Northern King Country—a prominent stronghold of Māori resistance to the Crown. In addition to being a geographically and logistically well-suited location for a power station, the example of Huntly represents how the MoW could use electric power infrastructures to ensure government power over land and people. Although I did not find specific records of public works officials choosing Huntly explicitly to displace the Māori, their actions and, at times, their ignorance of the importance of the place to the Māori highlights a potent colonial legacy. As historians like Jarrod Hore have argued, placemaking projects like settlement or photography have been part of the settler-colonial arsenal in Waikato since the 1860s. Situating a new power station to service the settler state, in one of its most productive coal towns, near one of the most politically powerful Māori communities, provided such an opportunity.[84]

Mahuta's critique demonstrates how NZED and the MoW treatment of the Tainui remained firmly tied to colonial practices throughout the construction of the Huntly Power Station. At the most fundamental level, the MoW still operated within the colonial frameworks set forth a century earlier. At the same time, Mahuta and others also used the criticisms to make broader arguments about how the Māori must align with other dissent groups. Otherwise, they risked being continually dismissed by a society that valued electric power utilities and the authority it grants the government over people and resources. He also recommended that Māori use development schemes that rely on and benefit Māori cooperation, if possible. In a report he produced later during the construction process, Mahuta made a note of the local rage from the town in one particularly potent piece from *The Huntly Press* on March 14, 1973: "It would be thought that the Maori community would be right behind the establishment of the power project—what with employment that will be provided, a large portion of the labour force required would be Maoris and they stand to gain immensely. . . . It is hoped that the protests now being voiced from Huntly West [a suburb of Huntly] will not spell doom to a project that could give Huntly district a badly needed 'shot in the arm' economically."[85] In addition to criticizing such patronizing remarks, which arose from established colonial rhetoric concerning Indigenous people in New Zealand, Mahuta explained in one of his reports that when local Māori expressed their concerns publicly, they received violent calls that included remarks such as, "What are you Māoris screaming about? Why are you concerned about your marae? We'll lose this station if you do not watch it."[86]

The supremacy of electric power generation is written into the well-defined yet ambiguous laws that govern public works in New Zealand. Land, they imply, like water, especially for producing electric power, can be taken and should be taken. Māori landowners faced the most incredible difficulties here because the state saw their land as lesser, owing to older relocation and confiscation schemes, not to mention outright racism. The state often argued, as they had done for over a century, that Māori land could be better used in productive service to the government. Furthermore, such a mentality encouraged the valuation of Māori land to decrease and therefore warrant minimal compensation. As documented by many scholars at the Centre for Māori Studies Research at the University of Waikato in the succeeding decades, these laws and policies created a sense of hopelessness toward the government's indiscriminate use of power to advance the Huntly Power Project.[87] Despite environmental protections and growing national concern for Māori rights, NZED and other departments wielded these laws to make the Huntly Power Station's construction and eventual opening an inevitability.

After the 1973 approval and the appeals, construction began in earnest. The MoW and NZED slowly brought around 2,262 workers in to build and run the plant, build new bridges to haul equipment, and construct housing.[88] The power plant switched on in 1983, and it is massive, featuring four massive boilers housed in twenty-story buildings. At first, coal powered the station, but gas also could be used after the pipeline from Taranaki was completed. The boilers sent steam at 538° Celsius to spin the turbines at 3,000 rpm and then exit via fifty-story chimneys. Huntly produced 953 megawatts of power at full capacity (expanded to 1,453 megawatts in 2007) and annually spewed 80,000 tons of fly ash.[89] The press, and often the corporate history of the station, viewed its completion as a turning point for the town and sometimes the North Island.[90] The aims of the earliest proponents of a 1,000-megawatt station in the North effectively had been achieved. The station produced more power than could be consumed and contributed considerably to the sprawling growth of Auckland, which would become home to more than half of New Zealand's population.

Although heralded as a success from the late 1980s well into the twenty-first century, the station received much criticism for environmental damage it caused—especially air and water pollution.[91] In the 2010s, growing pressure from the environmental movement and electric power supply authorities suggested that the power plant be decommissioned. In 2012, Genesis Energy, one of New Zealand's largest electric power companies, switched off one of the four boilers and has since decommissioned another.[92] Genesis Energy's website, which includes basic information about significant stations like Huntly, and even registration for a guided tour through the Huntly Power Station, says, "The iconic, orange-tipped power station has been producing energy for New Zealand for more than 40 years. Huntly is critical to the country's energy security and will play an interesting role in the future as we move to a low carbon future."[93] Energy security remains central to the plant's operation. Since the 2010s, New Zealand's reliance on renewables has increased considerably; today, around 82 percent of the country's electric power is produced by renewables.[94] Huntly's operators increasingly consider the plant a backup when drought and climate render renewables unusable. One of their boilers is testing biomass as an alternative fuel source.[95] The Huntly Power Station will remain a staple of electric power in New Zealand as a backup power source, especially as the country aims to reach its zero-emissions promises by 2030.[96]

In 1865, two years before Ferdinand von Hochstetter's report on New Zealand, William Stanley Jevons published a work titled *The Coal Question: An Inquiry Concerning the Progress of the Nation, and the Probable Exhaustion of Our Coal-Mines.*[97] He argued that the insatiable rate at which the British Empire consumed coal might explain their power and success but also doom them to a future in which they could never reach such prosperity: "Coal in truth stands not beside but entirely above all other commodities. It is the material energy of the country—the universal aid—the factor in everything we do. With coal almost any feat is possible or easy."[98] Beyond casting coal as a critical resource, Jevons exposes many things in this text about British anxieties, those within Great Britain and throughout its empire, about the future of resources that many used to further territorial gain and promote new power sources, like electricity. Although Jevons wrote of English industries and resources, his words reflect how people, indeed many of the people that colonized and later governed New Zealand, saw control over energy resources as a marker of thriving industry and society. In regional papers, journalists and community members echoed the country's excitement for electric power and the possibilities they believed it brought, while also acknowledging that risks existed. This is illustrated in this line from the *Waikato Times*: "[The people of Waikato] want the power station to be established but not on any terms, and any price."[99] Unfortunately, the risk and harm fell disproportionately on the Tainui. The drive to build the Huntly Power Station consumed those in power, including Mayor Michael Minogue of Hamilton City, just quoted. In 1977, George Gair, the minister of electricity, declared that despite the concerns raised by Mahuta on behalf of the Tainui, "destiny" brought the marae and the power station together.[100]

The construction of Huntly Power Station, beyond building a plant reliant on extractive industries, also refined the policies of the MoW (renamed the Ministry of Works and Development in 1974) to ensure centralized authority over resources. New Zealand's electrification is a product of Britain's anxiety about energy resources and colonial power. New Zealand's government, even after it became a dominion and achieved additional sovereignty following World War II, maintained and utilized the legislative systems carried over from earlier regimes to secure a way forward to produce more electric power, further encoding colonialism onto its electric grid. The decision of some community members and those on planning bodies behind the Huntly Power Station to diminish Māori culture and presume the superiority of their visions for New Zealand established a consistent line with earlier electric power boosters' language and public works law. New Zealand, and the Māori, were to submit to the vision held by the state, which valued the creation of a self-sufficient settler state in the Pacific.

# CHAPTER 5

# "A PEBBLE IN YOUR SHOE"

## GEOTHERMAL POWER AND SETTLER-COLONIALISM

On December 3, 2020, Rawiri Waititi and Debbie Ngarewa-Packer addressed Parliament as coleaders of Te Pāti Māori, or the Māori Party. Since his 2016 defection from the Labour Party, Waititi has strongly spoken against and challenged efforts in Parliament to silence Māori voices in Aotearoa New Zealand. He has, for instance, focused on prison reform, attire rules in Parliament, and COVID-19 protocol, and he has experienced both solidarity and pushback from his settler colleagues.

In the opening address, Waititi evoked a potent analogy that worked not only to describe how he understood his position and responsibility in Parliament but also as a metaphor for how Māori in the late twentieth and early twenty-first centuries challenged the settler logic of New Zealand's electric power infrastructure: as a pebble in the shoe. He said, "I will ensure our unapologetic Māori voice will be heard and that our Māori cloak is felt and is present in every piece of legislation and bill passed in this house. . . . You know what it feels like to have a pebble in your shoe? That will be my job here. A constant, annoying [*sic*] to those holding onto the colonial ways, a reminder and change agent for the recognition of our kahu Māori."[1] Like a "pebble in the shoe," the history of the Tūaropaki-controlled geothermal power stations on the North Island affords us the chance to stop and adjust how we talk about electrification in New Zealand. Although the trustees of these power stations did not iterate explicit designs to challenge the colonial patterns within the grid, their efforts nonetheless showcase resistance to settler designs in the history of New Zealand's electrification.

Throughout this history, especially in the case of Huntly, Māori frustrated the efforts of the public works bureaucracies that sought to electrify New Zealand their way. Even where their specific stories remain unknown, settlers often viewed the presence of Māori as a barrier. For centuries, the Māori have resisted settler colonialism through war, legislation, and community. Many Māori continue to resent and resist the invasion of Aotearoa. However, Māori are, of course, more than resistance. They are part of Aotearoa New Zealand. Many Māori identified and continue to identify as New Zealanders, and as recent historians have emphasized, Māori and settlers built New Zealand and defined New Zealander culture together.[2] Māori participated in both world wars, which gave some of them a greater sense of belonging because the government returned and granted Māori veterans new lands. Furthermore, the deployment of people from New Zealand as part of the larger Australian and New Zealand Army Corps forces in both wars forged a more connected sense of national identity through the shared trials of war at home and abroad.[3]

Unfortunately, as is true in so many of the archives preserving the British empire's history, the historical record obscures Māori contributions to New Zealand's electrification. Despite this lapse in scholarship, Māori were users and designers who participated in electrifying the country. When discussing electric power, we historians must acknowledge Māori resistance to and usage of electric power systems. In doing so, we will uncover a more complete history that sheds light on the colonial forces that shaped, maintain, and make exploitative the electric power infrastructures of New Zealand, while also drawing attention to the disruptions of that exploitation. This chapter recasts the colonization of the Māori in the process of electrification and then explores how one group of Māori electrified their lands to challenge colonial systems.

From the beginning of New Zealand's transition to electric power, settlers designed and used electric power systems in service to their colonial mission. For settlers, miners, and politicians, electric power provided the means to exploit the frontier, efficiently distribute resources, and facilitate the creation of a modern state. Although public works bureaucracies designed these infrastructures to enhance New Zealand's self-sufficiency, they also intended for them to establish the dominance of British settlers over the Māori. Settlers integrated electrification into colonial processes, such as communication, transportation, and resource extraction. As a result, writing through the history of New Zealand's electrification sometimes feels like an onslaught of victories for settler projects in which colonial institutions outpaced and outmaneuvered Indigenous peoples or reconciliatory settler energy projects. However, framing electrification as a process that can be lost or won misses the fact that the impact of electrification cannot be measured in such terms. The contours of New Zealand's colonial divisions cannot be so categorized. It's not just about settler victories or Māori defeats, nor is it just about more or less power or more productivity; rather, electrification shapes space and alters the planet.

Understanding and criticizing the colonial power structures within New Zealand's electrification history requires giving due attention to Māori actors and highlighting their narratives whenever possible. In New Zealand and elsewhere, the majority of electrical history presents electrical technologies and the requisite knowledge to electrify as primarily Western technological diffusion.[4] Histories of both British colonialism and electrification paint electrification as a settler success. According to New Zealand newspaper archives and Parliamentary records, the early stages of electrification there fortified the accomplishments of the Colonial Office. Only settler successes like lighting, mining, power supply, and trams remained visible, because they served the colonial mission. As Diana Montaño shows in her exploration of Mexico's electrification under the Díaz regime,[5] these perspectives on electrification privilege the success and dominance of settlers and imply the necessity of colonialism to make it happen. But even though Māori narratives in Reefton, Bullendale, Lake Coleridge, and Huntly convey tales of removal, absence, and conformity to Anglo-settler requirements, Māori played far from a passive role in the colonial projects that built Reefton and the Phoenix Mine through trade and land negotiations. Many Māori took out mining licenses, panned for gold during the rushes, and lived in mining communities.[6] At Lake Coleridge, Māori set up the agricultural industries and trading posts that helped establish Christchurch as an urban center.[7] Māori and settlers both advocated for and against the construction of the Huntly Power Station. In the latter half of the twentieth century, Māori built electric power production facilities and used that power to challenge colonial structures in New Zealand; reclaim land; and protect their heritage, environment, and future.

Today, New Zealanders cite the recent success of the country's geothermal energy infrastructure to promote climate change awareness, environmental preservation, and sustainability. New Zealand began pursuing geothermal energy in the 1950s as a viable solution to weaknesses in its existing generation methods, and has become a global leader in developing the associated technologies. In August 2023, the Ministry of Business, Innovation, and Employment celebrated its third year of growth in their renewable energy sector. The report claimed that 87 percent of the country's electric power is produced using renewable energy, about 60 percent of which is hydropower, 25 percent geothermal, and the remainder a combination of wind, solar, and biomass.[8] Unlike the electrical enterprises at Reefton, the Phoenix Mine, Lake Coleridge, or Huntly, many (though not all) of the country's geothermal power stations are rooted in energy practices arising from Māori culture. Māori investment, persistence, and environmental worldview in reclaiming Aotearoa New Zealand's resources, such as those at the Mōkai Geothermal Fields, have worked to expose the colonial policies built into New Zealand's electric power infrastructure while crafting space for Māori participation in New Zealand's local and global economic initiatives.

Māori utilization of Land Trusts and their community-conscious production of electric power provides a chance to explore the role of electrification in anticolonial resistance and explore colonialism's legacy in New Zealand's infrastructures. Māori Trusts have engaged in technopolitical legislation to create an energy policy that serves Māori peoples and accords with their sense of proper land use. Even as settlers used electricity to construct their colonial state, following the establishment of the Dominion and increased local settler authority, Māori used it as a powerful tool to deconstruct colonial infrastructures. Māori leaders and business owners, such as the Tūaropaki Trust, examined in this chapter, used and continued expanding geothermal electric power production to reclaim *Tino rangatiratanga* (absolute sovereignty), establish a diverse Māori economy, and develop a sustainable energy future for Aotearoa New Zealand. Building on a century of Māori land reclamation efforts, the Tūaropaki Power Company's investment in geothermal energy, sustainable farming, and social justice represents an active refutation of the colonial practices of early electric power systems in New Zealand.

Geothermal energy was deeply rooted in New Zealand's history long before the invasion of Europeans. Māori, especially the Te Arawa, a major confederation of iwi and hapū with shared ancestral ties, incorporated geothermal resources into their cosmogony and daily life on Aotearoa. For instance, Ngātoroirangi, a prominent tohunga (priest) from Māori legend known for being on the voyage from Hawaiki, the homeland of Māori ancestors, used geothermal energy to establish a home for his people.[9] During his exploration of Te Ika-a-Māui (Aotearoa's North Island), Ngātoroirangi became trapped on Mount Tongariro, dying of cold. He called out to his sisters, Kuiwai and Haungaroa, to bring him heat. Upon hearing his cries, they sent their spiritual being, in the form of fire, from Hawaiki through the Earth to warm Ngātoroirangi, leaving behind geysers, hot pools, and volcanoes throughout Maketū (Bay of Plenty), especially around Lake Taupō.

For centuries, Te Arawa settled and thrived in much of the area around the Taupō Volcanic Zone, including Lake Taupō and Lake Rotorua.[10] They used, and many still use, geothermal steam and water for cooking and heating.[11] In 1993, in front of the Waitangi Tribunal, Hāmuera Mitchell, a Te Arawa elder of the Ngāti Whakaue, recalled the relationship his community held with geothermal resources:

> That's how Ngāti Whakaue was in those days, and this place is 100% geothermal. All along the lakeshore was geothermal . . . we would go fishing for morihana . . . we had no clothes (no pakehas around in those days aye), and we would catch a few and tie them on a string, we would be shivering then we would dig a hole in the mea [thing] by the lake and the water come up and we'd be warm. If we wanted to cook our fish straight away, we'd dig another hole, and cook our breakfast. That's how we were in the old days, we lived it we spoke it we sang it and we slept in it.[12]

Such relationships existed around Te Ika-a-Māui for centuries.

By 1800, the European invasion of Te Ika-a-Māui began, particularly by the British. Trade, settlement, and resource exploitation increased, especially after the Treaty of Waitangi (1840). Even after the signing, colonial officials such as Donald McLean (1820–1877), who became one of the most influential settlers to handle Māori and settler relations during this period, wondered at the uses of the baths in his diaries on October 26, 1845: "The most remarkable features in this place are the immense number of Hot Springs in the very centre of the Pah [*sic*] rendering it dangerous to walk about after dark. It appears to be a complete subterraneous boiler of volcanic steam. These springs are not unattended by danger to the inhabitants, sometimes breaking out in fresh places and scalding those near to death. They are of different temperatures, some sufficiently hot for cooking food whilst others make agreeable baths."[13] In this passage and elsewhere in his diaries McLean concerned himself with evaluating potential hostilities and opportunities in uncharted spaces. His dismissal of Indigenous awareness of potential hazards typifies settler commentaries of the day, as does his focus on the resources.

During the nineteenth century, settlers capitalized on Māori geothermal resources in the tourist and health industry. In 1845, Robert Graham (1820–1885) founded the first settler-geothermal business in New Zealand, a spa at the Waiwera Geothermal Fields.[14] Throughout the 1870s, newspapers advertised the hot springs at Waiwera and elsewhere, promising that the springs "border on the miraculous."[15] In 1878, Graham, fluent in Māori, helped to solve a land dispute in the Taupō Volcanic Zone at Maketu. Te Arawa leaders granted him some land during this exchange. Historians debate about how he obtained so much territory and whether the leaders indeed granted him ownership. Nonetheless, he became the sole proprietor of the Ohinemutu Hotel, the Terrace Hotel at Wairoa, and 1,700 hectares between Huka Falls and the Wairakei Geyser Valley.[16] Besides jump-starting the geothermal tourism industry, Graham's land acquisition led to increased dialogue between settlers and Te Arawa, who had traditionally resisted land sale.[17] Eventually, the government and Te Arawa agreed to establish a township to facilitate the building and working of a large sanatorium. In parallel with these land acquisitions, the government passed the Thermal Spring Districts Act of 1881, which formally placed all thermal springs under government control because only the government could purchase the land. Even though these exchanges led to a thriving tourist industry that in some ways celebrated Māori culture and promoted cooperation, it continued colonial patterns of land acquisition that ultimately disenfranchised Māori.[18]

Following the signing of the Treaty of Waitangi, many Māori lost control of the land they owned through purchase, unfair land deals, and outright seizure in the aftermath of the New Zealand Wars (1845–1872).[19] Much of the acquisition took place through the enactment and amending of the Native Lands Act 1862, which also established the Native Land Court. The act, while still manipulative of Māori practices of land tenure, did represent an experiment in such tenure in the British empire and was unique to New Zealand.[20] The government passed these laws to individualize Māori land tenure in keeping with English common law. Traditionally, Māori land was communal; therefore, such legislation fragmented Māori authority, as it split their land ownership.

In many land acquisition cases, the New Zealand government spurned the Treaty of Waitangi.[21] In 1877, for instance, during *Wi Parata v. Bishop of Wellington*, Chief Justice Sir James Prendergast (1826–1921) ruled that land claims made by the Ngāti Toa against the Anglican Church were a "simple nullity."[22] Though some Māori political and social movements gained power during the late nineteenth and early twentieth centuries, the New Zealand Parliament and Colonial Office did little to negotiate meaningfully. Despite Māori efforts to participate in New Zealand politics, such as advocating for the inclusion of Māori representatives in Parliament, the settler government continued to inhibit their political influence. For instance, they forced Māori to deal directly with the Crown through appeals and petitions, rather than directly purchasing land or negotiating with local offices or among themselves.

As they continued to lose land during the nineteenth century, many Māori fought for ways to retain their lands and create infrastructures that could bind Māori land interests behind a single administration. As mentioned in chapter 4, during the 1850s, in response to unfair land losses on the North Island, many Māori iwi formed the Kīngitanga, or Māori King Movement, a system in which they selected a leader to represent them like a European monarch and negotiate with the Crown.[23] During the 1860s, Māori and settlers alike sought political means to ease relations during the New Zealand War. In 1867, Parliament passed the Māori Representation Act and added Māori seats to the legislative body, although these seats did not give them equal representation to settlers. Despite underrepresentation, Māori politicians and community leaders continued to advocate for changes to oppressive settler policies, including unsuccessful movements to redraw the electorate based on tribal regions and create a Māori Parliament (the Kotahitanga Movement).[24]

During the twentieth century, Māori land tenure fell into to the following categories:

- individual ownership, in which Māori acquired or purchased land in accordance with settler systems;
- multiple ownership, in which many parcels were owned by Māori iwi, whānau (extended Māori family group), or hapū in which owners collectively decided land use;
- Māori Land Court titles, in which the court decided whether to grant title based on ancestry and Māori understands of land tenure;
- Government Confiscation and Reserves, in which the government granted Māori confiscated land and established settlements and communities through government grants;
- Leasehold land, in which Māori Land could be leased to provide a source of income while Māori retained ownership; and lastly,
- Māori Land Trusts, which granted trusts the ability to manage and administer land in accordance with the trust's culture, history, and understandings of the environment.

The establishment of Land Trusts, such as the Waikato mentioned in the previous chapter, represents one of the most influential and longstanding Māori methods of regaining cultural and political authority.

Māori established businesses that reclaimed Māori land and began creating business models to serve Māori workers, farmers, and education. In 1929, the first reclaiming of land using government funding happened under the Māori Land Development Scheme, designed by Paratene Ngata (1849–1924) and Sir Āpirana Ngata (1874–1950), be9 cause they established businesses that provided work, financing, and materials for the Māori community.[25] Even though this new scheme continued to operate on structural terms of place-naming and land ownership, it did provide government funding to Māori landowners to develop land, effectively creating Land Trusts and allowing Māori to form agricultural incorporations. Āpirana Ngata strongly supported the expansion of sheep and dairy farming, popular trades among the Ngāti Porou. This Māori iwi, to which Āpirana Ngata belonged, is located on the northeastern coast of the Te Ika-a-Māui. After the 1890s, the iwi played a role in Māori revitalization, forming incorporations such as the Ngāti-Porou Dairy Company and Ruatōria Factory to generate Māori employment and income.[26] From 1925 until the 1950s, the Ruatōria-based dairy and factory gained fame for the Nāti-branded butter it produced. Māori ran the venture and the financing, which included buying herds and working the facilities. According to a 1952 review of the company, the dairy was "staffed and managed entirely by Māoris, and 90 percent of its cream supply comes from farms under Māori ownership or management."[27] Such practices became a standard model for other Māori Land Trusts. Even though some details of the trusts' operations, such as the Ngāti Porou Dairy, are open to researchers, many trusts maintain strict control of their business plans and records. Therefore, data on the success of the early trusts is limited. Nevertheless, we can note that Māori-sourced industries built to circumvent settler dominance reinforced anticolonial movements in New Zealand.

The Native Lands Act 1873 had created fragmented ownership and sharing of land for many Māori under colonial law because it disestablished communal titles. During the following decades, Māori land holdings decreased dramatically. In 1920, the Native Trustee Act created the position of Native Trustee along with the Native Trustee Office.[28] The Department of State charged this office with care and improvement of Māori land holdings. Historians and Māori, while highlighting some of the rural development of land the office accomplished under plans like Ngata's, have maintained that it largely continued to alienate Māori. For example, the act empowered the government to reallocate land deemed as surplus into Crown ownership or private hands. In 1947, the Native Department became the Department of Māori Affairs and in 1953, following the passage of the Māori Trustee Act, the Native Trustee position was replaced by that of the Māori Trustee. This person did not have to be Māori, but they were no longer supposed to be under the control of Crown and were charged more explicitly with protecting the Māori rights and interests.

During these legislative shifts in the early twentieth century and their subsequent revisions in the late twentieth and early twenty-first centuries, the Native Department and later Department of Māori Affairs (now Te Puni Kōkiri or Ministry of Māori Development) used the collective ownership of Māori land as a tool for socioeconomic development. Māori Land Trusts allow for the retention, use, and development of land in accordance with Māori values and economic aspirations. Early trusts, such as the East Coast Māori Trust, were often imposed by the Crown to manage debts, but they frequently facilitated land alienation rather than protection. Over time, reforms sought to center Māori agency, creating structures like whānau trusts, focused on familial ownership, and incorporations, designed for economic activity. The Land Trust system provided the legislative authority for the trusts, such as the Tūaropaki Trust, to act as individuals over blocks of land while representing many Māori landholders. Even though land trusts might be understood as an assimilationist scheme enforced by settler ideas of land title, the trusts comparatively offered Māori more leeway. The Māori and Crown practices regarding land title remain incredibly complex and have received extensive treatment by settler and Māori scholars, including Richard Boast, Richard Hill, and Dion Tuuta. There are also a number of excellent studies on Māori trusts, such as Marama Muru-Lanning's *Tupuna Awa*, Stella Coram's *Extinguishing Title*, and Richard Hill's *State Authority* or *Maori and the State*.[29] Collectively, they demonstrate how scholars with and without direct links to Māori communities can expose Māori political agency, explain settler tactics, and highlight the significance of the trusts in New Zealand's history.

## MĀORI AND EARLY TWENTIETH-CENTURY ELECTRIFICATION

Following the success of Lake Coleridge, New Zealand's electric grid expanded across the North and South Islands. As of 1916, of the 49,776 Māori in New Zealand, 1,900 lived on Te Waipounamu, the South Island. According to the census, these Māori lived almost exclusively in settler cities or nearby settlements. The implementation of electric power in the Canterbury Region almost certainly affected Māori lands by redirecting waters, constructing power lines, and lighting shared spaces. However, determining whether these Māori had access to domestic power is impossible because we have no census data on Māori dwellings.[30] Te Ika-a-Māui, the North Island, is where the most Māori chose or were forced to live following the New Zealand Wars.[31] Even though electricity flowed to Te Ika-a-Māui's major cities, it did not reach most Māori, except for city-dwelling hāwhekaihe, or people of mixed Māori and non-Māori ancestry.[32] There were exceptions, such as Parihaka, Te Ika-a-Māui's largest Māori settlement. Other North Island major hydroelectric facilities, like the Arapuni Power Station (1929), which would supply much of the power for Auckland, were built during this period. By the time the North Island's electrification began to scale, electric power legislation would have made it difficult for any town, Māori or otherwise, to generate its own power without state oversight. With few exceptions, such as Waitomo, Waihou, or Matamata, Māori communities would not receive electric power until the 1950s, often along with larger or adjacent settler communities.[33] Māori Land Trusts possibly may have devised small-scale generators, but the archival documents reviewed during this project have not yielded evidence for them.

Even settlers lamented the conditions of the Māori suburb of Orakei, including Auckland resident George Graham, who argued, "Given the security of residential tenure, sanitation, electric light, and the other amenities of modern life Orakei village and its dwellers could be transformed into a happy community—instead of being, as things stand, a reproach to the authorities responsible."[34] To be sure, his language contained settler assumptions about the future and assimilation of Māori, but it nonetheless highlights lack of access. After all, Europeans had come into New Zealand touting (in the *Evening Post*'s view) that "wonder of wonders, the electric cable, which brings us within a few hours communication of the rest of the world."[35] Despite such noble proclamations and efforts to electrify mining towns and industrial cities, the same government initiatives did not exist for Māori communities. Public works budgets and plans can be found for Rotorua and other communities that fit into New Zealand's burgeoning tourist industry, but many other communities, such as those around Huntly, did not receive electric power until the 1930s. Even then, it becomes difficult to track where electric power hookups extended and who used it. There are numerous reports of accidents, such an unfortunate youth participating in a duck hunt near Waahi Pā who was electrocuted after climbing a pylon, but details about people's familiarity with electric power remains scant, at least using traditional textual sources.[36] As the previous chapters demonstrate, the state wielded electric power works to expand their political economy; by the same token, the lack of access to electric power disadvantaged communities.[37]

During the mid-twentieth century, due to a growing Māori population and a shortage of settler workers, the Ministry of Works (MoW) employed Māori laborers to update and expand New Zealand's electric power system. Many of these projects formed the basis of the grid that would connect the country during the 1950s; as mentioned earlier, Māori, along with laborers from around the world, built the primary power lines that connected the various local power boards. Even though Māori did not write the legislation or draft the engineering plans, their labor helped build the grid.[38] In doing so, they created their ideas about what it meant to make electric power structures. Bill Hiku, one of the Māori linemen, fondly remembers Māori workers racing up pylons and hanging from wires unsecured as a show of masculinity and physical prowess.[39] Other linemen also celebrated their role in state-building by arguing that working on the line gangs gave them a sense of belonging to the nation because they worked alongside settlers to build public works.[40]

The rapid expansion of the electric power grid in this period exemplified one of the many public works initiatives that took advantage of smaller Māori properties. The Crown had long used the ambiguity surrounding Māori freehold land, or land returned to or granted to Māori by the Crown, to facilitate developing land for public works, especially after the significant reforms of the 1870s but certainly earlier too.[41] Despite the cessation of large-scale government land acquisition and the success of Māori land trusts during the late nineteenth and early twentieth centuries, public works authorities sought their lands because only the government could sell the land, which artificially lowered the value.[42]

Following the successes of hydroelectric expansion during the 1930s and 1940s, the Public Works Department saw geothermal energy as one way to supplement New Zealand's growing demand for electricity, particularly for industrial manufacturing and export industries. Besides using geothermal wells to produce electric power, many in New Zealand had already incorporated them into heating systems. Throughout these decades, New Zealanders, especially around Rotorua, bored into geothermal wells to tap into "Nature's Boiler."[43] Even though the system used a different method of power production, not hydro or fossil fuels, geothermally powered systems still relied on the central government's ability to use the land to expand the country's productivity, in line with a century-old settler ideology. The government codified it in the Geothermal Steam Act 1952: "Geothermal steam includes steam, water, water vapour, and every kind of gas, and every mixture of all or any of them, that has been heated by the natural heat of the earth . . . Subject to the provisions of this Act, the sole right to take, use, and apply geothermal steam for the purpose of generating electricity shall vest in the Crown."[44] The government, first under the guidance of the MoW and then the New Zealand Electricity Department (NZED), built the country's first four geothermal stations at Wairakei (1958), Kawerau (1989 and 1996), and Ōhaaki (1989), where they sold or leased Māori-owned land under compulsory land acquisition for "public works."[45]

The construction of the Wairakei Power Station during the 1950s and 1960s makes plain how the government forcefully pursued every means of expanding electric power at any cost. W. H. Goosman (1890–1969), who, as minister of works, oversaw the first investigations into Wairakei's generating potential, said at an interview in the Wairakei Hotel in 1950, "I intend to press this survey as rapidly as possible, so that we may be able to reach a decision on the practicability of the suggestion within a year or 18 months."[46] The Wairakei Power Station began producing power in 1964 and had a peak output of 192 megawatts (MW), compared to the 34.5 MW Coleridge generated in 1930 and the 953 MW Huntly would produce in the 1970s.[47]

Today, energy experts typically classify geothermal energy as renewable energy. While we can now operate geothermal systems in a sustainable, environmentally conscious manner, many of these early systems used exploitative frameworks of electric power production focused more on tapping new resources than on preserving them. During this period, geothermal power production relied on the production of "dry steam" facilities.[48] Wells had to produce steam at 150°C or higher so that vapor could be transferred directly up a shaft to turn turbines, collected in a condenser, and sent back in the wells to be cooled. While dry steam remains one of the most efficient methods of geothermal power production, this method caused subsidence, or the downward settling, of land due to the removal of underground liquids.[49] Reversal processes exist in some cases, but most wells will disappear, as those at Wairakei did. In yet another form of exploitation, in 1961 the government also compulsorily acquired the Ōrākei Kōrako geothermal system in the Waikato Region to develop the Waikato River's hydroelectric Ōhakuri Dam. Approximately three-quarters of the area's geysers, sinter springs, and other geothermal features were submerged, as was the site of the primary marae (meeting place) of the Ngāti Tahu, an iwi traditionally tied to the territory between the Rotorua lakes and Lake Taupō and a hapū of Ngāti Tūwharetoa. Since then, groups within the iwi have either bought back or had land returned, although the dam remains and its lake still covers many of the geothermal features. Many of the trusts in the region have subsequently leased the central area of these features to a tourism concessionaire.

Reckless settler exploitation of the Māori geothermal sites for tourism and energy echoed the practices of earlier colonization because settlers utilized them for profit and without consideration for Māori. Even though reparations have been paid for some losses, many sites suffered irreparable damage. The Wairakei and Ōhaaki power stations caused the extinction and depletion of geothermal geysers and springs highly valued by Māori. The impact of Wairakei's construction and operation, ranging from collapsed wells and mercury leaching to the destruction of Māori heritage, continue to be studied. It still remains a part of the grid under the control of Contact Energy Ltd. The historically adverse effects on valued geothermal features, initially handled by the NZED, are now managed through cooperation between Contact Energy and Te Kupenga Charitable Trust, set up by members of the Ngāti Tuwharetoa in 1996. The trust established a Māori tourism operation, Wairakei Terraces, using bore water supplied by Contact Energy to create and maintain artificial geothermal features and bathing pools. The trust has also set up Nectar, a Māori training organization with three training facilities.[50] Its key focus concerning education and training has been to create opportunities and employment for Māori in education, health, and tourism.

From the installation of the Reefton system forward, some Māori and settlers recognized that hydroelectric systems ruined water sources for agriculture, drinking, and culturally significant purposes. However, for most settlers, the progress achieved through electricity outweighed these problems. During the 1950s, settler environmental activists began to seriously protest the damage inflicted by these dams, namely at the Manapōuri and Huntly facilities. This movement resulted in the 1970s in conservationist legislation that curbed pollution.[51] Despite successes like this, overall New Zealand's power grid expansion led to an electrical system that increasingly abused New Zealand's environment, reinforced colonial visions about the country's productivity, and enforced a stronger centralized settler government.

During the 1970s, in what is sometimes termed the "Māori Renaissance," Māori, frustrated with unexamined land claims, vied for a new legal mechanism. After decades of controversy, protests, marches, and the establishment of the Māori Affairs Amendment Act 1974 by the Honorable Matiu Rata (1934–1997), New Zealand established the Waitangi Tribunal, Te Rōpū Whakamana I te Tiriti o Waitangi, in 1975.[52] Parliament placed the Tribunal in charge of investigating Māori claims of breaches of the Treaty of Waitangi by the Crown and, by extension, the New Zealand Government. Some scholars, such as Andrew Sharp and Bill Oliver, argue that the institution of the Tribunal represented the tame option because it merely averted Māori protests rather than confronting deep-seated colonial malfeasance.[53] Initially, the standing commission of inquiry only heard cases for breaches occurring from 1975 forward. By 1980, only seven claims had been investigated. The Labour Party then expanded the Tribunal's timeline back to 1840.[54] Still, the Tribunal could only make recommendations to the government; they did not have any actual judicial authority. Although flawed, the Tribunal's attempts at reconciliation encouraged anticolonial movements. Some argue that it was and remains an essential milestone in correcting grave injustices against Indigenous peoples.[55]

On February 1, 1977, the Whātāpaka Marae and subtribes of the Waikato made a claim through the Waitangi Tribunal against the NZED for its plans to build the Auckland Thermal No. 1 Power Station. The station, like Huntly, proposed the injection of water back into Manukau Harbor. However, unlike Huntly, the claims halted the construction. Auckland was never built, although NZED did pursue other thermal stations in the region, such as the New Plymouth Power Station. However, many of the issues the Tribunal used in their decision are directly linked with ecological uniqueness or the shellfish economy. Certainly, the Māori there benefited from that economy, but their cultural resources and heritage did not.

After decades of increasing national authority in New Zealand's electric power generation, a group of Māori decided to fund and de-

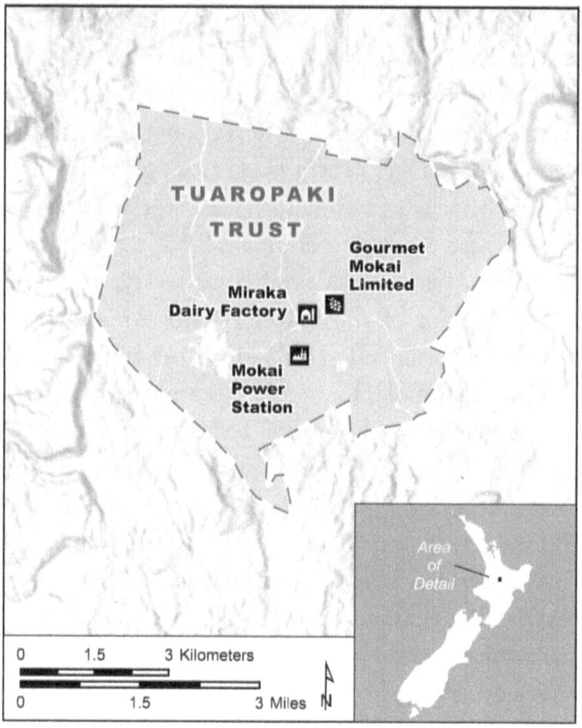


FIGURE 5.1.
Tuaropaki Trust.

velop a geothermal electric power scheme that disrupted the electric power production practices embedded in the public works infrastructure since the 1880s. They called it the Tūaropaki Power Company.

## THE TŪAROPAKI POWER COMPANY

The Tūaropaki Trust is a collection of multiple-owned lands traced to the descendants of seven Mōkai hapū: the Ngāti Parekaawa, Ngāti Kohera, Ngāti Wairangi, Ngāti Whaita, Ngāti Moekino, Ngāti Haa, and Ngāti Tarakaiahi (figure 5.1).[56] Initially, the trust served to further Māori agricultural interests but grew to incorporate electric power production, education services, and telecommunications. Even though the focus here is on the Tūaropaki Trust, the power company's success resulted from collaboration between many Māori trusts, companies, and individuals. This partnership resulted from centuries of resistance to settler governance and increased Māori integration and revisions to settler bureaucracies. For instance, in 1914, Māori land owners leveraged royalties for the use of their land by the Taupo Totara Timber Company around the small community of Mōkai (which, because of its proximity to the town, later became the name of the geothermal power station) to gain access to the Putaruru–Mōkai railway.[57]

As noted earlier about other trusts, the Tūaropaki Trust does not share or provide public access to their records. The primary documentation used in this analysis comes from inf ion published in New Zealand's press, the trust's open access documents, Waitangi Tribunal reports, and other scholars who have been given access. For this project, I obtained and reviewed that body of scholarship and archival material from primarily online databases. Beginning in 1951, area landowners amalgamated to develop a single pastoral farm. The following year they formed the Tūaropaki Development Scheme and succeeded in obtaining approval of the loan from the government. The collective received financing from the Department of Māori Affairs, which regulated the farms until the trust repaid the loan in 1975. During the 1960s, the trust coordinated a comprehensive development program under the department to expand farms known as the Tūaropaki Land Development Scheme.[58] In 1982, the owners repaid the loans and interest, and the government returned all of the lands to a board of trustees, which formally established the Tūaropaki Trust. Operations included cattle and sheep farms at first, but owners gradually diversified the stock.

In 1993, the Waitangi Tribunal recognized Tūaropaki as an Ahu Whenua Trust acting under the provisions of that year's Te Ture Whenua Māori Act (Māori Land Act). As part of more extensive Waitangi Tribunal reforms, this act reaffirmed the Treaty of Waitangi. It stipulated that "that land is *taonga tuku iho* of special significance to Māori people," and "they [Māori] facilitate the occupation, development, and utilization of that land for the benefit of its owners, their *whānau*, and their hapu." Taonga tuku iho translates as "cultural heritage" or "birthright," but literally means "treasure gifted down." Whānau means "extended family" but can also refer to all Māori.[59] This act, along with decades of negotiating and resisting colonial regimes, provided the foundation for and catalyzed the mission of the Tūaropaki Power Company on the land just east of Lake Taupō.

Prior to the 1950s, Māori and settlers used the region for timber, dry stock farming, and dairy farming. The owners of Tūaropaki reoriented the focus from those resources to the energy potential trapped in the thermal vents underground. The Taupō Volcanic Zone (TVZ) is one of the most active geothermal sites in the world. During the

1950s, the zone drew the attention of the MoW, riding on the perceived success of major electrification projects like Meremere (1958) and the speculative talks of a station near Huntly, as they moved away from planning more hydropower stations. Journalists used the MoW's speculations about the TVZ to diminish Māori title and highlight electric power potential, as the following excerpt shows: "On the surface, Mokai is a quiet place. A marae, a school and a handful of worn, bare houses are the ghost of a timber town which in the 1930s had four mills and 5000 people. . . . But below this land, 30 km northwest of Taupō, lies the hottest geothermal field in New Zealand and a potentially valuable source of electrical power." Other than mention of the "swimming holes for the locals' enjoyment," settler perceptions of Māori knowledge about the utility of geothermal power trivialized their long history and knowledge of their environment.[60] As had happened at other wells like Wairakei, the Department of Scientific and Industrial Research (DSIR) began drilling to test the resources. But they proved expensive, and many believed the area did not provide the correct incentives—scenic views, recreation, and tourism—to attract and house the people needed to build and manage a major geothermal power station. Also, MoW officials, despite the promise of more versatile generation options, worried about potential environmental and water rights backlash because of projects like Huntly and other geothermal developments around Rotorua. Ultimately, the government did not pursue the development of wells.

However, in 1983, during their survey of the Mōkai geothermal fields along with corresponding scientists from geothermal initiatives around the world, the DSIR noted that the entire field might generate 150 megawatt electrical (MWe), or 25 MWe for each of six injectable wells.[61] The wells are some of the hottest in New Zealand, regularly reaching temperatures of 325°C, whereas the majority operate between 140°C and 220°C.[62] In 1987, during explorations of the TVZ's geothermal potential, the government, and eventually the Ministry of Energy's assets as the deregulated Electricity Corporation of New Zealand (ECNZ), continued to market electricity to expand New Zealand's economy, enlarge urban spaces, and maintain its settler population. The ECNZ was formed that year from the remnants of the older public works infrastructure, as a way to consolidate the underregulated local supply authorities under the Ministry of Energy, the body that took over after the Electricity Department reforms of the 1960s. It was with this body that the Tūaropaki Trustees negotiated as they formed their power company.

During the government's exploration of the TVZ, the Tūaropaki Trustees opposed and successfully prevented the government's development of the region. In 1994, they acquired resource consents to build and run a geothermal power station as the Tūaropaki Power Company Limited. After the Department of Māori Affairs secured Tūaropaki's interests—meaning the trustees purchased the land from the Crown—the trust began to develop the Mōkai geothermal fields in 1996, with plans to build a power station. The trustees agreed that Tūaropaki's success hinged upon when they, not the government, acted as owners and developers. During the mid-1990s, some trustees suggested that they outsource or lease the land to provide more income for their existing agricultural initiatives, but the trust ultimately decided that they should remain in control of their interests as owners and developers.

The next step involved selecting how to harness the steam for electric power. As electricity developers had done on many of the country's prior projects, the Tūaropaki elected to work with engineers and consultants from abroad. Since they had fought to obtain the right to develop and construct the plant on their terms, Tūaropaki trustees elected an engineering and equipment company aligned with their interests. According to Sam Joesph Andrews, a senior trustee at the time of this decision, "We have a terrific opportunity at Mokai to create and build the economic power of our people by applying the concepts of continuous sustainability to our commercial operations. Māori look upon geothermal resources as taonga which must be respected and used in ways which benefit Māori now and in the future."[63]

Tūaropaki selected Ormat Pacific as the contractor for the project. By the 1980s and 1990s, Ormat had grown from a small turbine manufacturer into a major renewable energy company specializing in solar power, geothermal, and retrofitting existing power plants to increase efficiency. Following the construction of their first geothermal power plant in 1984, the Wabuska Power Station in Nevada (now part of the large and divisive Dixie Meadows Geothermal Project), Ormat became a global leader in geothermal plant engineering and construction. In 1992, they developed the 50 MW Leyte Optimization Project in the Philippines, and in 1998 they started plans for the 140 MW Olkaria III complex in Kenya.[64]

During Tūaropaki's early investigations, they learned that Ormat had gained a foothold in New Zealand by having constructed the 55 MW Poihipi Power Station (1996), controlled by Mercury Energy; the 24 MW Rotokawa Power Station (1997), controlled by Transalta; and the 10 MW Ngawha Power Station (1998), controlled by Top Energy.[65] The state-owned electric companies developed and controlled each of these Taupō-based stations. The Tūaropaki trustee's aims aligned most closely with the work Ormat had done at Rotokawa. Most notably, Ormat had built the station quickly using a modular design; reused exploration wells the Crown drilled decades prior; and had minimal environmental impact through Ormat's use of reinjection, steam trapping, and low-rise condensers. According to Ormat and engineers working on other geothermal projects in New Zealand, the trustees chose Ormat for these reasons.[66] Access to the original negotiations, conversations, or contracts between Tūaropaki and Ormat remains difficult. Nevertheless, two major reasons the Tūaropaki Trustees chose Ormat remain clear—their patented reinjection technology and their willingness to follow the Tūaropaki's demands.

The construction of the plant was itself an exercise in Māori innovation and pragmatism on the chosen 2,700 hectares of land, known as Tūaropaki E. They determined Geothermal Combined Cycle technology (GCC) was the most appropriate method of power production in a 60 MW plant, unlike systems that draw large amounts of water from other sources, which require the water to be cooled, treated with hazardous chemicals, and then disposed of. In a GCC system, a backpressure steam turbine produces steam, which a binary plant vaporizer subsequently condenses, offering more power. The liquid product of this process is then reinjected back into the ground.[67] GCC technology is more modular and requires less environmentally damaging and time-consuming construction. The binary turbines require only simple low-level foundations because they are smaller, more efficient, and do not rely on higher steam temperature or volume. Also, the turbines do not need an attached condenser, so the condenser units sit on a lower foundation. Because of the smaller facilities and absence of a large cooling tower, this plant style does not create a significant visual impact, an essential factor in Tūaropaki's decision to use this design.

The government, which still held authority over electric power generation projects, granted Tūaropaki permission to build in February 1998. The turbine generator arrived in February 1999. Tūaropaki Power Company and Mighty River Power, one of the three state-owned generating companies, commissioned the first plant at the end of that year.

## DISRUPTING CURRENTS

The new technology and the promise of commercial gain by a Māori trust might appear to function as a contortion of the same settler tactics deployed by the projects discussed earlier. And indeed, sustainability certainly has been appropriated by power project planners and companies under capitalism to mask extraction and profit. However, the intention of the commercial existence of the power station challenges such assumptions. As mentioned by Andrews, the Tūaropaki saw geothermal resources as taonga, referring in that context to something culturally treasured or prized; it is Māori and it is to be preserved for Māori now and in the future. Taonga has evolved over the centuries for Māori communities engaging with settler epistemologies, like the scientific categorization of animals, but also for energy resources. The pursuit of profit to benefit people culturally and legislatively oppressed for two centuries reveals how Māori used a settler-colonial framing of electric power production to undo some settler legacies.

In this way, the Tūaropaki Power Company challenged and transformed some of the colonial motivations behind electric power production in Aotearoa New Zealand. As Dominion residents, settler citizens and politicians employed colonial ideas about New Zealand's purpose and how it should generate electric power. Even after the Electricity Act 2003 dissolved ECNZ to create the Electricity Commission, and later the Electricity Authority (2010), to regulate electric power with less Ministry of Energy oversight, the central aim of the new entity was to turn a profit for power companies, preserve New Zealand's energy self-sufficiency, and register all power providers under a single regulatory authority.[68] The rebranding of New Zealand's electric power marketplace, even with their occasional attention to environmental sustainability, continues to prioritize the expansion of electric power production at the expense of New Zealand's environment and people. The preference for profit is made most evident by the country's recent debates concerning drilling for more natural gas to sell overseas and generate electricity.[69]

In many ways, earlier settler electrification projects and Tūaropaki share the goal of self-sufficiency. Yet there is a crucial difference in what the respective groups have deemed necessary to accomplish independence. Settler systems relied on hegemonic practices like resource extraction and land theft. By contrast, the Tūaropaki Company used profits and energy from the plant to build and power new sustainable industries that employ and educate Māori. Māori efforts return political power to Māori, contribute to the Māori community's economic improvement, and use energy in ways consistent with Māori perspectives on resource use. Unlike settler electric power facilities, bent on constant expansion and unsustainable industrial profit, Tūaropaki's primary goal has been to "act as a beacon of hope and prosperity for our people."[70] They use electricity production to foster Māori employment, education, financial aid, and environmental sustainability in New Zealand and beyond. As an example of the latter, the trust funded a massive research effort into the nitrogen and phosphorus concentrations in the Mississippi River Basin between 1980 and 2018.[71]

Tūaropaki elders hold and promote the belief that "if you look after the land, the land will look after you."[72] For the Tūaropaki Māori, they and settlers exist in a constitutive partnership to sustainably produce energy in New Zealand. This restructuring of electric power production has repurposed some colonial legacies and granted Māori more authority, like land trusts and electric power bureaucracy, to develop a system that benefits Māori, settlers, and the environment. Sustainability is an integral part of the trust's guiding principles. However, when the Tūaropaki Trust evokes sustainability, it is only partially aligned with settlers' understanding of the concept. Although both Māori and settler power companies, policymakers, and communities find value in energy independence, environmental preservation, and green initiatives, Māori sustainability directives counteract colonial power structures in the energy grid that work to discount their existence and understanding of energy. Beginning with Mōkai 1 Power Station, activated in 2000, the trust has operated from the philosophy that neglecting the environment will prevent long-term survival at both the macro and micro levels. In keeping with this philosophy, they have established measures to repurpose waste from all of their facilities.

The Tūaropaki Trust's geothermal development has promoted environmental awareness and wellbeing throughout New Zealand. Of

FIGURE 5.2. Gourmet Mokai, September 2, 2014. United Nations Photo, Evan Schneider UN 7227575.

their many projects, the facilitation of Māori farming is one of the most publicly celebrated accomplishments. Using geothermal steam as a heat source, the Tūaropaki Trust launched Gourmet Mokai for growing sustainable year-round crops for local consumption and export. In December 2002, Tūaropaki entered a joint venture as a 25 percent shareholder to establish . Gourmet Paprika Ltd., an Auckland-based greenhouse horticulture company, holds a 50 percent share, and the other 25 percent is held by Hauhungaroa Partnership. Hauhungaroa is a 200.48-hectare property on the western side of Lake Taupō. The partnership is among many small trusts on that land. Gourmet Mokai agreed that the structure would sit on Tūaropaki land, and the Taupō fields could provide the necessary heat. In 2006 the trust expanded the 5.5-hectare climate-controlled glasshouse to 11.7 hectares (figure 5.2). The facility directors sought to keep in step with the Tūaropaki's commitment to longevity and low environmental impact. The greenhouses use hydroponics to grow individual plants in a soilless medium linked to centralized water and fertilizing systems. The facility produces capsicum (bell peppers), Truss tomatoes, blueberries, and other rotating crops.

Following the success of Gourmet Mokai, in 2010 the trust formed a partnership with Wairarapapa Moana, Waipapa 9 Trust, Hauhungaroa Partnership, Tauhara Moana Trust, Pouakani Trust, Te Tumu Paeroa, and Te Awahohonu to establish Miraka, a dairy processing facility, on Tūaropaki land. The company created a permanent Māori-owned organization. Tūaropaki agreed to host the dairy processing plant because the land's central location benefits Māori dairies in the region and harnesses geothermal power. Construction began on the plant in late 2010 and ended in December 2011. The facility includes all the components needed to produce, pack, and warehouse powdered milk. The entire plant is powered by renewable steam and electricity provided by the Tūaropaki Power Company, the first such facility in the world. The roughly 60 gigajoules of waste heat produced daily are processed and recycled at the facility. As of 2017, the Miraka plant could process milk from 50,000 cows (210 million liters annually) into 32,000 tons of whole milk powder. Another Māori Trustee, Global Dairy Network, has sold the product to foodstuff industries in China, the United States, Malaysia, Taiwan, and Australia.[73]

In 2012, Tūaropaki opened the Ngaire George Sustainability Centre, named after a longtime trust chair, Ngaire George, also chairperson of the Mōkai Marae Trust, a subsidiary trust of Tūaropaki, to expand their sustainability model.[74] Wastewater and other materials are taken from Miraka, the glasshouses, and a number of the area's dairies and processed by worms to produce vermicompost. Vermicomposting decomposes waste rapidly, and geothermal steam keeps the mixture at optimal temperatures. As of 2018, the center has 4.5 kilometers of worm rows dedicated to composting waste, a native plant nursery that produces 200,000 plants annually, a riparian planting program for all Tūaropaki lands, and a research center dedicated to preservation. During the first two decades of the twenty-first century, these projects served the trust's mission to continually reduce the environmental impact of its facilities. It removed the energy, outsourcing, and logistics costs associated with agricultural waste removal, which produces a nutrient-dense compound that Tūaropaki's industries can recycle.

The Tūaropaki Company's establishment of Gourmet Mokai, Miraka Dairy, and Ngaire George Sustainability Centre demonstrates how the Māori sought to use their geothermal resources to reclaim lands lost to colonial exploitation, provide work for Māori, and create a sustainable export industry for New Zealand. As of 2017, 2,400 owners receive economic dividends and social benefits from these industries, not to mention the New Zealanders that use its electricity or the global customers of their dairy industries. In 1982, the trust made $4 million from their agricultural facilities; by 2016 they had made nearly $16 million and employed 366 Māori in the Waikato region.[75] On January 28, 2019, they signed an agreement with Japan's Obayashi Corporation to develop a "zero-carbon" hydrogen production facility powered by Tūaropaki's geothermal power.[76] These successes were made possible by Māori innovation and their commitment to sustainability, founded in producing electric power using geothermal energy.

Using geothermal energy, the Tūaropaki Power Company has made tremendous contributions to geothermal technology and altered the makeup of New Zealand's power production methods. In particular, they have demonstrated the utility of using direct heat from geothermal sources on an industrial scale. The expected growth of electric power demand exceeds the capabilities of existing and planned wells. Directly using the steam to power manufacture or provide heat thus reduces the demand for electric power. Regarding environmental impact, from the beginning the station's has been minimal. Under regular operation, the geothermal fluid remains contained from production to reinjection. This system existed in the first 55 MW station built in 1999, the 40 MW station added on in 2005, and another 2007 addition that brought the plant's output to 100 MW. From the method of power production to the facility's design to its income distribution, the Tūaropaki Power Company, much like the land trust system, challenged while having to exist within the colonial underpinnings of New Zealand's electric power network, prioritizing productivity, self-sufficiency, and centralized state power.

It is worth noting how Tūaropaki concerns and New Zealand's settler electrification infrastructure have aligned, even if for different reasons. To view Tūaropaki's actions as separate from these structures is to look at this case study with the same settler gaze that insisted the Māori could not function in the modern Anglo-world. But as scholars like Maria Bargh and Frédéric Sautet argue, the Māori function in the same economy as the settler structures, not outside of it.[77] Even if their reasons are based in cultural understanding of resources, they also exist in a diverse economy in which electric power is a commodity. My use of *diverse economy* here borrows from economists' argument that the local economies of the geothermal works are not mere emulations of traditional capitalism but a more explicit expression of Māori authority. Thus, the full story of New Zealand's electrification is one with an inescapable colonial legacy, although electricity itself is by no means intrinsically hegemonic. On Tūaropaki's website, the trustees claim that their mission is to "use wisdom, knowledge, and science to address current and future challenges." They structured their mission around five guiding principles:

1. We will act as a beacon of hope and prosperity for our people.
2. Protect and advance the interests of our people.
3. Be at one with our cultural land and resource.
4. Help build the capability of our people through positive support and encouragement.
5. Look after the land, and the land will look after you.[78]

These principles and the mission apply to Tūaropaki interests that extend beyond electric power production into arenas like food production, education, and vocational training. Nonetheless, they should be unpacked as part of their efforts to build a geothermal power station. As Māori trustees, they decouple the electricity production from the settler state authorities and emphasize culture, community, and environmental sustainability as part of their plan to generate power. Yet simultaneously, they connected that station to the grid, built primarily by and for the settler state. In 2023, the 113 MW produced at Mōkai travel 22 kilometers and connect to Transpower, one of the country's largest state-owned enterprises, which made $127 million in profits during 2023.[79] This is not a critique of the Tūaropaki Trust; rather, it is a reassertion of the prescience of recognizing settler-colonial infrastructures in New Zealand's electric power systems.

## POSITIONING MĀORI POWER PRODUCTION IN THE HISTORICAL RECORD

Much of the records available concerning Tūaropaki work from 1996 forward use settler and Western ideas and rhetoric. Most of the published information on Tūaropaki and the Mōkai power stations comes from scientific perspectives, in journals like *Geothermics*, or from conference proceedings, like INAGA, the Indonesian Geothermal Association.[80] Such sources focus on technological achievements, economic impacts, and financial partnerships, but still provide useful insight into the Tūaropaki's decision-making process. Although much of the Tūaropaki's work may be described using Western scientific terminology, they did not uncritically repeat or assimilate this knowledge. They made it their own by profiting from the high efficiency of geothermal power while insisting that the importance of efficiency extends beyond profit to managing decreasing energy availability worldwide, combating economic inequality created by New Zealand's existing grid, and adhering to their stewardship of the land.[81]

The Tūaropaki Power Company's story and ongoing projects expose the absence and active removal of Māori from the history of electrification and encourage others to grant Māori a place in energy history. As indicated in the Introduction, meaningful research on the absence of Māori must be approached in concert with Kaupapa Māori scholarship, a critical and anticolonial base that asserts an understanding of New Zealand's history as a partnership between the settlers and Māori.[82] Following scholars like Linda Tuhiwai Smith and Graham Smith, who use the empowering outcomes model, I have aimed to produce research that might serve the Māori people and expose exploitative systems for those unfamiliar with New Zealand's past.[83] To guide the process and work, research practices must be structured to center Indigenous peoples' concerns.

A growing number of scholars argue for the acceptance of knowledge production that exists outside the bounds of accepted Western understandings of electric power generation. Yet even to insist that Māori must appear in these narratives as much as settlers do reaffirms settler notions of legitimacy. Inquiring into the Māori role in history requires careful treatment of the limited historical sources and recognition of the ways that Māori have at times been left out of the historical record, but also acknowledgment of the scholars, such as Lachy Paterson and Angela Wanhalla, who highlight the Māori voices preserved in the written record of the nineteenth century.[84] The limits of the archive make clear the power of erasure and the longevity of colonialism, and perhaps historians who use these archives need to focus more on the possibilities of locating and doing good Māori history. Much of the textual postinvasion Māori history covered in this chapter comes from legal documentation concerning land, but some information about recent Māori electrification history is unavailable because of the Māori decision to keep it private. This work therefore relies heavily on non-Māori sources and authors.[85]

Social scientists have often looked to Indigenous languages to more fairly and accurately report on how Indigenous peoples understand the natural world. In this way, they hope to avoid using Western categories (like "resource management"), which may be inappropriate or fail to capture the complexity of Indigenous belief systems. In New Zealand's environmental legislation and energy policy since the 1990s, many settlers and Māori have adopted the Te Reo Māori term kaitiakitanga, which means "guardianship," in New Zealand's legislation and energy policy. As part of the Waitangi Tribunal and the state's growing recognition of Māori culture in the 1970s and 1980s, the term entered public discourse to promote intercultural cooperation. Currently, politicians, energy policymakers, and activists, both settlers and Māori, use the term in environmental discourse aimed at conservation and sustainability. Yet the use of the term may cause more problems than it solves. Merata Kawhuru and others argue that the term should not be translated as "guardianship" nor be limited in scope to preserving the environment.[86] On the one hand, the term has had a significant impact on the country's environmental legislation, especially the Resource Management Act 1991, which effectively brought all of New Zealand's resource-use regimes under one administration that operated under the principle of sustainability.[87] On the other hand, this usage does not represent the breadth of meaning of the term within Māori cosmogony. Māori interpretations of kaitiakitanga weave through the relationship between human, material, and nonmaterial elements, not just ideas about resource management or production.[88] Also, different iwi do not necessarily share a clear definition; using it further disenfranchises some of the many cultures that make up the Māori. It is beyond the scope of this chapter to explore the expansive and changing meanings of the term. Rather than add to the uncritical usage that threatens to oversimplify how Māori understand and aim to improve the world, I have tried to explain as clearly as possible in conventional language what the administrators of the Tūaropaki Trust respect and aim for in their geothermal project.

In 2018, the Ngāti Tahu and Contact Energy received an award from the United States Energy Association for their efforts.[89] The World Energy Council's Trilemma rating index placed New Zealand as eighth in the world for its energy security, sustainability, and equity, with many wealthier European nations scoring higher. However, the council has noted that while other countries have better infrastructures in place, the model set forth by Māori-led geothermal power projects offers one of the most promising for improving energy equity around the world. Energy equity is increasingly relevant to island populations threatened by climate change.[90]

In New Zealand, the Tūaropaki Power Company demonstrated how Māori could reclaim their lands; promote environmental sustainability; and transform colonial electric power production models that deplete resources, are concerned primarily with profit, and emphasize centralization. Amid mounting concerns over climate change, geothermal energy has proven a practical aid for other forms of alternative energy by supplementing wind and solar farms during off-hours. The Tūaropaki Power Company development points to how electric power companies might adopt energy systems that can change, rather than replicate, colonial technological infrastructure.

Geothermal electricity production continues to expand in New Zealand, and the Tūaropaki Power Company has served as a model for subsequent Māori and settler geothermal schemes. On October 31, 2018, Eastern Group, a New Zealand–based company, opened the largest geothermal power plant in New Zealand to date, the 25 MW Te Ahi O Maui Plant.[91] Eastern Group partnered with the Kawerau Ahu Whenua Trust, contracted Ormat Technologies, and made sustainability their mission. It is a striking parallel to Tūaropaki and similar to a number of other recent developments, such as Taheke 8 (2010), Tauhara North No. 2 (2010), and Tikitere Geothermal Company (2011).

The success of Māori geothermal power ventures and related industries has generated international interest in promoting similar projects worldwide and in New Zealand, the most recent being the Olkaria Power Station in Kenya. Since the 1980s, the Kenya Electricity Generating Company (KenGen) and the Maasai peoples have clashed about the development of geothermal fields. In 2017, with the encouragement of Power Africa, the Maasai and KenGen began a series of exchange missions with Māori at the Ōhaaki Geothermal Power Station in Waikato to learn how the power companies and Māori Trusts managed their relationships. Using Māori values to guide its operation, much like the Tūaropaki Trust, Māori operate Ōhaaki. The company provides mechanisms for underprivileged youth to enter the energy sector, maintains a sacred pool, and provides flood mitigation structures. Following many exchanges, KenGen and the Maasai have begun to work out ways to establish a measure to share land ownership, observe Maasai cultural ties to the region, and develop sustainable energy for Kenya.[92]

Like many of the electric power production milestones seen in previous chapters, the Tūaropaki Power Company completed several "firsts" in the electrification of New Zealand. However, like other Māori Land Trusts before them, Tūaropaki explicitly committed to Māori-led electric power production and shifted ideas about resources and energy consumption—namely, the idea that electric power is meant to serve the country's economic expansion or its energy security. The Tūaropaki, because of the embeddedness of settler-colonial infrastructures, developed a means to generate electric power, profit from its sale, and redistribute the profits among trustees. Their development and innovation provides historians with excellent cause to explore alternate pathways to electric power production and forces historians of New Zealand to interrogate the long-lasting exploitative technological structures of New Zealand that continue to thrive.

## CONCLUSION

# ELECTRIFICATION IS NOT A FINISHED PROCESS

> The ravaged beauty God alone could plan!
> Bitter the thought: "Is this the price we pay—
> The price for progress—beauty swept away?"
>
> **—William Pember Reeves, "The Passing of the Forest," 1898**

In 1962, biologist-activist J. T. Salmon published *Heritage Destroyed,* a book protesting environmental degradation and promoting reform in Aotearoa New Zealand's preservation system. He used the country's hydroelectric systems to highlight his frustrations: "In New Zealand the projects to develop our river systems for electrical power are being executed as engineering works only, without any thought for conservation principles, for landscape design, or for the preservation or enhancement of scenic attractions. . . . The likely economic loss is not weighed against the probable economic gain of cheap electricity."[1] Environmental activists and scholars in New Zealand cite Salmon's book as one of the many publications that signaled the beginning of a decade-long reorganization of the National Parks Authority and a broader environmental movement in New Zealand.[2]

Internationally, power companies and activists have recognized

New Zealand as a sustainable and green country.[3] Numerous international organizations commend it for its commitment to renewable energy. As mentioned earlier, most of its renewable power comes from hydroelectric dams. Despite the benefit of a reduced carbon footprint, the country's green status depends on centuries of colonial practices. Large hydroelectric infrastructures have required land transformation to provide massive amounts of electric power. Even if they result in less net pollution, they reinforce exploitation that does not produce truly sustainable energy. To grapple with these issues, New Zealand will have to reconstruct its electric power generation practices.

At the heart of Salmon's book was the construction of the Manapōuri and Aratiatia power stations. Much like the resistance to the Huntly Power Station, protesters resisting these projects argued that the facilities would dramatically harm the country's environment (specifically Lake Manapōuri and the Aratiatia Rapids). In contrast, the New Zealand Electricity Department and the aluminum smelting company Consolidated Zinc imagined that the project might generate enough electric power to unlock the "electro-industrial" potential of New Zealand's aluminum resources,[4] an all-too-familiar story of extractive and exploitative politics that would resonate with environmental movements that challenged state power over nature. Many New Zealanders agreed that the development of such massive hydroelectric infrastructures did not adequately weigh the consequences of electric power production for New Zealand's people or natural resources. In fact, the public works officials in charge of electrical infrastructures had never done so. They and the country's extractive industries simply pursued electrification to exert control over New Zealand's people and resources.

Despite the protests, the Public Works Department completed the Manapōuri Hydroelectric Plant (800 MW) in 1971 to power an aluminum smelting facility and a large swathe of the Southern Island Transmission Network. The issues raised by the Manapōuri controversy, such as connecting New Zealand to the global economy, extracting the country's mineral wealth, exercising state power, and maintaining government control over the land, have been central to the country's electrification from the beginning. The Manapōuri plant is not an exception. It is the rule. Except for some recent restructuring, the foundations of New Zealand's electric power infrastructures remain complicit in the settler-colonial practices that lit Reefton, powered the Phoenix Mine, built Lake Coleridge, and circumvented environmental policy at Huntly—and also spurred Māori geothermal developments that actively combated them.

By examining these transformative moments of New Zealand's history, we can understand settler-colonialism's influence on modern systems. The electrification of Reefton, a mining outpost on the West Coast of the Southern Island, illustrates how the Colonial Office and settlers used electric lighting to envision a "new" New Zealand. Before the official colonization of the islands (1840), the British saw New Zealand as the epitome of a colonial frontier. In Reefton, the introduction of electric lighting proved British technology's supremacy, resonated with settler ideas about self-sufficiency, and promised to expand the island's productive potential. It was not an anomaly or technological marvel ahead of its time. Electric power in Reefton became another exercise in British settler-colonialism to aid in extracting resources, encouraging settlement, and controlling land. Electric light and power rhetorically appealed to a town filled with speculative investors and people in business seeking profit from the mines and miners, a dedicated resource frontier. The city serves as a starting point for highlighting how electric systems in New Zealand fulfilled colonial interests as much as they did local interests in lighting and power. Production stakeholders in electric power and lighting intended to expand production and tame the frontier in the British Empire's name.

Similarly, the Phoenix Gold Mine's electrification highlights the motive behind establishing hydroelectric power systems in New Zealand: making the country more economically productive. Mine owners and city officials did not pursue hydroelectric structures because they were novel or preserved the environment. They destroyed existing environments and made the landscape suitable only for extractive industries and power production. Electrification promised to increase the mine's profits, conquer and utilize natural resources, and highlight British technological superiority. Hydroelectric power at the Phoenix Mine transformed New Zealand from a materially exploitable region on the periphery into a model colony ready to deploy new technological systems. As a cornerstone in the history of the country's hydroelectric systems, the Phoenix Mine provides a means for understanding the colonial motives built into subsequent hydropower projects in New Zealand.

Just as electric power systems had fulfilled colonial imaginaries of a self-reliant colony, the expansion of and demand for electric power gave some officials cause to justify increased central government. The establishment of New Zealand's first national electric system epitomizes hydroimperialism in New Zealand because its construction hinged upon the repurposing of Indigenous land, the establishment of national electric-power regulations, and support from a state agency, the Public Works Department. The centralization of electric power production behind a central authority was one of the many infrastructure developments that strengthened the cause for New Zealand's new national government. Furthermore, the dam demonstrates how Dominion, even with the pretense of independence, remained firmly within Britain's settler-colonialism, especially in New Zealand. Rather than being New Zealand's first step toward a modern electrical grid, the Lake Coleridge Power Station adapted British colonialism to provide the physical infrastructure that justified a national government and that government's control over energy resources.

The effort to construct the Huntly Power Station highlights how people could use, or ignore, environmental policy to gain land and expand the grid. The national government's plans to control energy resources shifted in the wake of concerns over energy predictability and the oil crisis, which laid the groundwork for the Public Works Department (eventually the Ministry of Works) to pursue fossil-fueled electric power production. In their desire to maintain control over energy resources, even during the vast expansion of environmental policy and Māori rights movements, the public works agencies treated people and land as little more than expendable barriers.

Even at the end of the twentieth century, when the state-run Electricity Corporation of New Zealand disbanded and reconsolidated into private power boards, the physical infrastructure enforced centralization, large-scale production, and exploitation of the land—colonialism. The process of reclamation that the Tūaropaki Trust endured evidences how those exploitative policies continued to thrive well into the 1900s. The mere existence of Tūaropaki initiatives to level unequal access showcase this point.

As with so many other stories of electrification, these narratives can lead to the conclusion that specific methods of power production are inevitable. None of them are. These practices developed because they benefited political systems that relied on exploitation for expansion and production. That is why paying attention to moments when

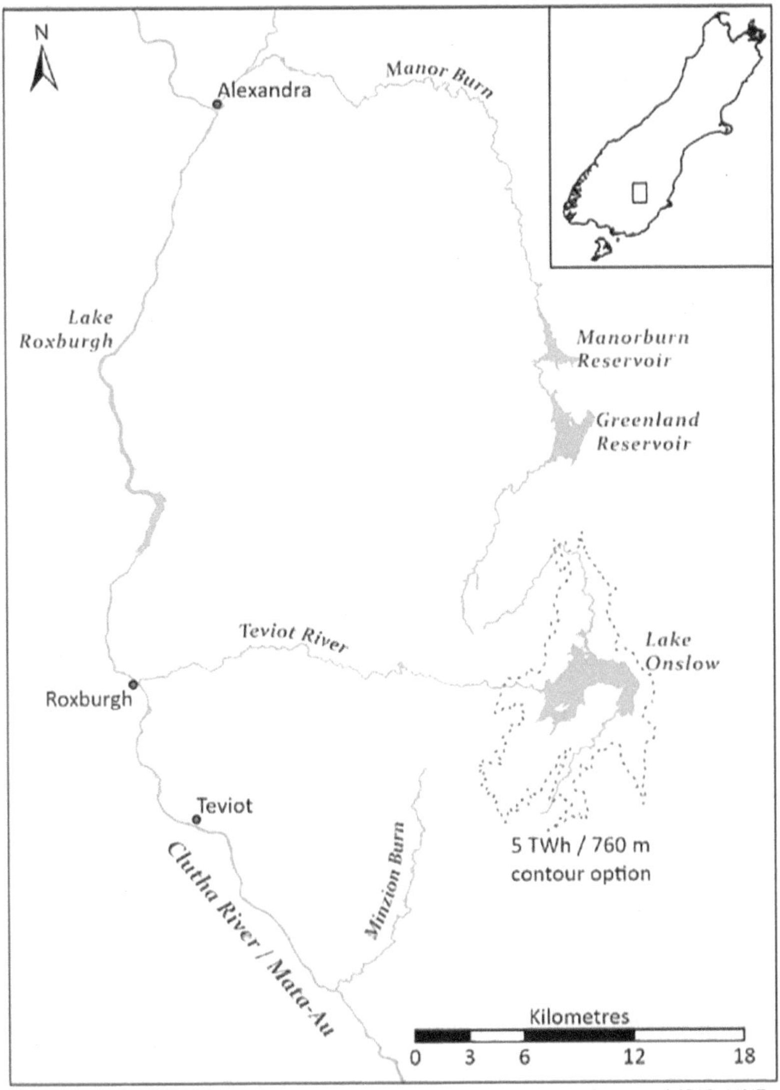


FIGURE C.1. Map of the Lake Onslow option as part of the New Zealand Battery Project. One of the options proposed 5 TWh dam that would be constructed at 760 meters above sea level. The dotted line illustrates approximately where the new shoreline would fall. *Cultural Values Statement: The Lake Onslow option for the New Zealand Battery Project*, prepared by Auhaka Ltd. for the New Zealand Ministry of Business, Innovation, and Employment, p. 5, accessed June 8, 2026, https://www.mbie.govt.nz/dmsdocument/28329-cultural-values-statement-the-lake-onslow-option-for-the-new-zealand-battery-project.

people disrupt those patterns is so important. Just as New Zealand cannot remove or ignore its settler-colonial history, it must acknowledge these colonial legacies built into its hydroelectric systems. Otherwise, they privilege exploitative energy infrastructures.

As we sit at the precipice of climate catastrophe, expanding inequity around energy access, and intensifying sociopolitical turmoil, not only must we confront the fact the electric power systems played key roles creating this reality; we also might also offer some path toward solutions. It is no coincidence that settler-colonial policy founded and continued to shape New Zealand's electrification. Electric power systems developed there because they afforded colonizers literal and figurative power. And although electric power need not continue to follow colonial paradigms, until we expose the exploitative realities of all electrification—yes, even renewable projects—it will do so in New Zealand and beyond.

Beginning in December 2020, New Zealand's government began Phase 1 of the "New Zealand Battery Project," a multiyear project to solve the "dry year" problem caused by the country's heavy reliance on hydropower without using more fossil fuels during years with less precipitation. The plan is part of the longer-term commitment to a "zero-carbon" future.[5] Much of the early focus has been on a pumped hydro-storage system at Lake Onslow, an artificial lake developed in the 1890s on the Teviot River in southwest Otago (figure C.1).[6] As mentioned by Prime Minister Ardern in the Introduction, the project would expand the lake, submerge land, cost NZD 15.7 billion, and provide an estimated half year of electric power storage. Like the work at Huntly, the government legally mandated environmental and social impact investigations, which have shown that the ecological losses and community impacts could be dire. At the conclusion of Phase 1 in February 2023, other locations were proposed and further explorations into Lake Onslow were extended. On December 3, 2023, the National Party's new government scrapped the Lake Onslow effort.

The new energy minister, Simeon Brown (who became famous in 2023 for advocating to remove Te Reo Māori from street signs), argued that "our decision to cancel [the] Lake Onslow [project], and our commitment to make it easier to consent wind, solar and geothermal energy projects, will give industry certainty in the direction the Government is heading and greater confidence to invest in more energy production in New Zealand."[7] Brown's claims recall scenes from prior electric schemes and government decisions that emphasized industry growth, investment, and government oversight. This decision and the government's justifications portend a troubling future for New Zealand's electrification that might continue to echo the settler-colonial idea of a successful New Zealand—one that privileges investment and industry. On the one hand, the decision to stop the project appears to cede victory to those with concerns for the environment and over cultural losses of rivers and plains. On the other, the outright dismissal of the project is not founded on a desire to protect the environment or Māori culture; it is about saving money and locating other ways to generate more power for industry. However, in early 2026, the Clutha Pumped Hydro Consortium, a collection of private investors from the energy sector, revived discussions about the project, citing grid vulnerabilities and a looming "dry year."[8] The Lake Onslow narrative fits the pattern established across the long history of New Zealand's electrification. While much remains to be seen about what will happen at Lake Onslow and with the Battery Project, the project's cancelation and reappearance matches New Zealand's pattern of electrification well. The ability to see electrification projects through has seldom depended on engineering feasibility, financial constraints, or the diverse interests of the population of Aotearoa New Zealand. What matters is whether those electrical infrastructures align with and support the New Zealand imagined by colonizers and their successors.

Clarity and transparency about the extractive reality of all energy systems is not a solution but rather a necessary step to limit exploitation. Just as no system that conducts electric current can stop the loss of energy, we cannot change the fact that energy production is extractive. Yet we still can insulate and diminish the loss of energy both during conduction and in the process of extraction. Electric power utilities can reconfigure the uneven impacts of power production. And perhaps more immediately, they can work to better understand how existing energy systems developed so they can be retooled. Electrification is not a finished process. Although exploitative power structures, such as the settler government of New Zealand, fundamentally shaped many of the electric power systems around the world, they are not inherently permanent.

In 1827, Georg Ohm solidified resistance as a key variable in the creation and measurement of a circuit, a relationship now known as

Ohm's Law. The law governs how electrical engineers design circuits, choose components for electrical systems, and manage the efficiency of electric power distribution. And while physics provides the best mechanisms for quantifying electrical phenomena, it does not always clarify how we interact with electric power. We are part of how electric power systems have to be understood. As I processed how best to bridge the difference in the ways scientists and humanists understand electric power, the term *resistance* remained constant throughout this project. Although Ohm may never have suggested it, perhaps resistance extends beyond physics and should include how we think about electric power systems historically and in the present. There are moments in the history of Aotearoa New Zealand and worldwide in which people changed the pattern of electrification or resisted the flow of the majority of the discourse. Perhaps those moments of resistance fail, or only affect a small piece of infrastructure, but they exist, shape other electric technology, and demonstrate other paths. Resistance is not only a fundamental variable of a circuit. It is a necessary piece of the process of electrification and a key to understanding the impact of the electrical systems that surround us.

## APPENDIX

# TIMELINES FOR THE GENERAL HISTORY OF AOTEAROA NEW ZEALAND AND ITS ELECTRIC POWER BUREAUCRACY

### GENERAL TIMELINE

| | |
|---|---|
| 1250–1300 | Māori Arrival |
| 1642 | Abel Tasman charts New Zealand |
| 1769 | James Cook spots New Zealand |
| 1790s | British settlement begins |
| 1839 | New Zealand Company |
| 1840 | Treaty of Waitangi/Te Tiriti o Waitangi |
| 1861 | Otago Gold Rush |
| 1870 | Immigration and Public Works Act |
| 1886 | Phoenix Mine builds an electric stamper |
| 1886 | Reefton Electric Light Company Founded |
| 1907 | New Zealand becomes a Dominion |
| 1914 | Lake Coleridge Power Station switched on |
| 1952 | Tuaropaki Trust Established |
| 1953 | Māori Affairs Act |
| 1983 | Huntly Power Station switched on |
| 2000 | Mōkai 1 Power Station switched on |

# ELECTRIC POWER BUREAUCRACY TIMELINE

| | |
|---|---|
| 1865 | Public Works Department |
| 1946 | State Hydro Electric Department |
| 1958 | State Electricity Department |
| 1977 | Ministry of Energy |
| 1986 | Electricity Corporation of New Zealand (ECNZ) |
| 1998 | Breakup of ECNZ |
| 2000 | Electricity Commission |
| 2010 | Electricity Authority |

# NOTES

## NOTE ON MĀORI LANGUAGE

1. For examples of the conversations and short history of this debate, see Stephen May, "New Zealand Is "Racist as F**k": Linguistic Racism and Te Reo Māori," *Ethnicities* 23, no. 5 (2023): 662–79.

2. Edgar A. Burns, "The Contested Shift to a Bicultural Understanding of Place Heritage in Aotearoa New Zealand," *International Journal of Heritage Studies* 29, no. 9 (2023): 988–1003.

3. This is by no means entirely agreed upon by Māori, Indigenous peoples, or scholars of Indigenous cultures in New Zealand or elsewhere. Should my decision have unintended interpretations, my hope is that fruitful conversation will be had about how we recognize and use indigenous language in scholarship. Gregory Younging, *Elements of Indigenous Style: A Guide for Writing by and About Indigenous Peoples* (Brush Education, 2025), 87–88.

4. Alison Jones, *This Pakeha Life: An Unsettled Memoir* (Bridget Williams, 2021), 6.

5. John C. Moorfield, *Te Whanake, Te Aka: Māori-English, English-Māori Dictionary and Index* (Pearson, 2005).

6. Ani Mikaere, "Racism in Contemporary Aotearoa: A Pakeha Prob-

lem," in *Colonizing Myths, Māori Realities: He Rukuruku Whakaaro* (Huia, 2011), 119.

7. Avril Bell, *Becoming Tanagata Tirititi: Working with Māori, Honouring the Treaty* (University of Auckland Press, 2024).

## PREFACE

1. For a more comprehensive exploration of the similarities of both spaces, see Harry Evison, *Te Wai Pounamu: A History of the Southern Māori During the European Colonization of New Zealand* (Aoraki Press, 1993) or *The Ngai Tahu Land Deeds: A Window on New Zealand History* (Te Rūnanga o Ngāi Tahu, 2006); Erwin C. Hargrove, *Prisoners of Myth: The Leadership of the Tennessee Valley Authority, 1933–1990* (Princeton University Press, 1994); Michael J. McDonald and John Muldowny, *TVA and the Dispossessed: The Resettlement of Population in the Norris Dam Area* (University of Tennessee Press, 2002).

## INTRODUCTION: ELECTRIFICATION AS NEW ZEALAND'S COLONIAL IMAGINARY

1. Jacinda Ardern, "100% Renewable Electricity Generation by 2030," Labour Party of New Zealand, September 10, 2020, https://web.archive.org/web/20200919023407/https://www.labour.org.nz/release-renewable-electricity-generation-2030.

2. Elijah Hill, "Wireless Electricity for the Masses Could Become a Reality Thanks to Kiwi Startup," *Stuff*, May 28, 2022.

3. Alexander Richter, "Geothermal Energy Provides a Great Opportunity for U.S. Indigenous Peoples," Think GeoEnergy, last accessed November 28, 2023, https://www.thinkgeoenergy.com/geothermal-energy-provides-a-great-opportunity-for-u-s-indigenous-peoples/.

4. Andre Needham, *Power Lines: Phoenix and the Making of the American Southwest* (Princeton University Press, 2014); Meiton Fredrik, *Electrical Palestine: Capital and Technology from Empire to Nation* (University of California Press, 2019). See also some other recent works in the history of electrification wrestling with these questions, such as Casey P. Cater, *Regenerating Dixie: Electric Energy and the Modern South* (University of Pittsburgh Press, 2019) and Diana Montaño, *Electrifying Mexico: Technology and the Transformation of the Modern City* (University of Texas Press, 2021).

5. Vaclav Smil, *Energy and Civilization: A History* (MIT Press, 2017); Richard Rhodes, *Energy: A Human History* (Simon & Schuster, 2018).

6. Erik van der Vleuten, "Understanding Network Societies: Two De-

cades of Large Technical Systems Studies," in *Networking Europe: Transnational Infrastructures and the Shaping of Europe, 1850–2000* (Science History Publications, 2006), 279–314.

7. Thomas Hughes, *Networks of Power: Electrification in Western Society, 1880–1930* (Johns Hopkins University Press, 1983).

8. Vaclav Smil, *Energy in World History* (Westview Press, 1994).

9. Andreas Malm, "Fleeing the Flowing Commons: Robert Thom, Water Reservoir Schemes, and the Shift to Steam Power in Early Nineteenth-Century Britain," *Environmental History* 19 (2014): 55–77.

10. Benjamin Sovacool, *Energy and Ethics: Justice and the Global Energy Challenge* (Palgrave Macmillan, 2013), 11–13.

11. Christopher Jones, *Routes of Power: Energy and Modern America* (Harvard University Press, 2016); Andreas Malm, *Fossil Capital: The Rise of Steam Power and the Roots of Global Warming* (Verso Books, 2016).

12. Andrew Curley, "Resources Is Just Another Word for Colonialism," in *The Routledge Handbook of Critical Resource Geography*, ed. Matthew Himley et al. (Routledge, 2021), 79–89.

13. Julie A. Cohn, *The Grid: Biography of an American Technology* (MIT Press, 2017); John L. Neufeld, *Selling Power: Economics, Policy, and Electric Utilities Before 1940* (University of Chicago Press, 2016); Paul Hirt, *The Wired Northwest: The History of Electric Power, 1870s–1970s* (University Press of Kansas, 2012).

14. David Nye, *Electrifying America: Social Meanings of a New Technology* (MIT Press, 1992); Nye, *American Illuminations: Urban Lighting 1800–1920* (MIT Press, 2022).

15. Crosbie Smith, *The Science of Energy: A Cultural History of Energy Physics in Victorian Britain* (University of Chicago Press, 1998); Jennifer Karns Alexander, *The Mantra of Efficiency: From Waterwheel to Social Control* (Johns Hopkins University Press, 2008); Barri Gold, *ThermoPoetics: Energy in Victorian Literature and Science* (MIT Press, 2010).

16. Gold, *ThermoPoetics*, 72; Arianna Turner, "Thing with Energy: Holism and the History of Energetics," Forum on Environmental History of Energy Transitions, *Environmental History* 24, no. 3 (2019): 482–89.

17. Iwan Rhys Morus, *Frankenstein's Children*, 184; Bruce Hunt, *The Maxwellians* (Cornell University Press, 2005), 100.

18. James Belich, *Replenishing the Earth: The Settler Revolution and the Rise of the Anglo-World, 1783–1939* (Oxford University Press, 2009).

19. Edwad Cavanagh and Lorenzo Veracini, *The Routledge Handbook of the History of Settler-Colonialism* (Routledge, 2017); Zak Cope and Imman-

uel Ness, eds., *The Oxford Handbook of Economic Imperialism* (Oxford University Press, 2022).

20. Lorenzo Veracini, *The World Turned Inside Out: Settler-Colonialism as a Political Idea* (Verso Books, 2021), 8–9.

21. For example, Rudolf Mrazek, *Engineers of Happy Land: Technology and Nationalism in a Colony* (Princeton University Press, 2002); Ben Marsden and Crosbie Smith, *Engineering an Empire: Technology, Science, and Culture, 1760–1914* (Palgrave Macmillan, 2003); William Storey, *Guns, Race, and Power in Colonial South Africa* (Cambridge University Press, 2008); Satia Priya, *Empire of Guns: The Violent Making of the Industrial Revolution* (Penguin, 2018); Elisabeth Koll, *Railroads and the Transformation of China* (Harvard University Press, 2019); On Barak, *Powering Empire: How Coal Made the Middle East and Sparked Global Carbonization* (University of California Press, 2020).

22. Duncan Bell, "Dissolving Distance: Technology, Space, and Empire in British Political Thought, 1770–1900," *Journal of Modern History* 77 (2007): 523–62.

23. Mary Louise Pratt, *Imperial Eyes: Travel Writing and* Transculturation (Routledge, 1992).

24. Daniel Headrick, *Tools of Empire: Technology and European Imperialism in the Nineteenth Century* (Oxford University Press, 1981); Headrick, *The Tentacles of Progress: Technology Transfer in the Age of Imperialism, 1850–1940* (Oxford University Press, 1988).

25. David Arnold, *Everyday Technology: Machines and the Making of India's Modernity* (University of Chicago Press, 2013); Arthur Asseraf, *Electric News in Colonial Algeria* (Oxford University Press, 2019).

26. James Belich, *The New Zealand Wars and the Victorian Interpreation of Racial Conflict* (Auckland University Press, 1986), 25; *Replenishing the Earth*, 108–9.

27. Charles Buller, "Systematic Colonization," Thursday, April 6, 1843, printed by John Murray; Edward Wakefield, *The Collected Works of Edward Gibbon Wakefield*, ed. Muriel F. Lloyd- Pritchard (Collins, 1968), 178–87.

28. Robert Pemberton, *The Happy Colony* (Saunders and Otley, 1854), 54.

29. Edward Wakefield, *A View of the Art of Colonization: With Present Reference to the British Empire: In Letters Between a Statesman and a Colonist* (John W. Parker, 1849), 44, 273, 497.

30. Although Japan did not owe Britain "allegiance," the British maintained input and some degree of control over the flow of electrical knowl-

edge. R. A. Buchanan, "The Diaspora of British Engineering," *Technology and Culture* 27, no. 3 (1986): 501–24, https://muse.jhu.edu/article/889526.

31. Miranda Johnson, "Burdens of Belonging: Indigeneity and the Re-Founding of Aotearoa New Zealand," *New Zealand Journal of History* 45, no. 1 (2011): 102–12; T. Boraman, "Indigeneity, Dissent, and Solidarity: Māori and Strikes in the Meat Industry in Aotearoa New Zealand During the Long 1970s," *International Review of Social History* 64, no. 1 (2019): 1–35.

32. Peter Gibbons, "Cultural Colonization and National Identity," *New Zealand Journal of History* 36, no. 1 (2002): 5–17.

33. David Hackett Fischer, *Fairness and Freedom: A History of Two Open Societies* (Oxford University Press, 2012)

34. Huhana Forsyth, "An Identity as Pākehā," *AlterNative* 14, no. 1 (2018): 73–80.

35. Tukana Mate Nepe, "E hao nei e tneei reanga: Te Toi Huarwea Tupuna" (master's thesis, University of Auckland, 1991), 226.

36. Manying Ip, *Being Maori Chinese: Mixed Identities* (Auckland University Press, 2008); Tony Ballantyne, *Entanglements of Empire: Missionaries, Maori, and the Question of the Body* (Duke University Press, 2014); Danny Keenan, *Te Whiti O Rongomai and the Resistance of Parihaka* (Huia Press, 2016); Felicity Barnes, *Selling Britishness: Commodity Culture, the Dominions, and Empire* (Auckland University Press, 2022).

37. Kate Darian-Smith et al., *Britishness Abroad: Transnational Movements and Imperial Cultures* (Melbourne University Press, 2007), 1–17; Kevin Grant et al., eds., *Beyond Sovereignty: Britain, Empire and Transnationalism, c. 1880–1950* (Palgrave Macmillan, 2007); Angela Woollacott, *Settler Societies in the Australian Colonies: Self-Government and Imperial Culture* (Oxford University Press, 2015), 1–2.

38. Terri A. Hasseler and Paula M. Krebs, "Losing Our Way After the Imperial Turn: Charting Academic Uses of the Postcolonial," in *After the Imperial Turn: Thinking with and Through the Nation*, ed. Antoinette Burton (Duke University Press, 2003).

39. Ballantyne, *Webs of Empire*, 11.

40. Richard Hill, *State Authority, Indigenous Autonomy: Crown-Maori Relations in New Zealand/Aotearoa 1900–1950* (Victoria University Press, 2004); Hill, *Maori and the State: Crown Maori Relations in New Zealand/Aotearoa, 1950–2000* (Victoria University Press, 2010); Atholl Andersen et al., *Tangata Whenua: A History* (Bridget Williams Books, 2014).

41. Felicity Barnes, "Rethinking Cultural Colonization: New Zealand's

Heritage and the Writing of New Zealand History," *New Zealand Journal of History* 57, no. 2 (2023): 1–22.

42. Ella Henry and Hone Pene, "*Kaupapa Māori*: Locating Indigenous Ontology, Epistemology, and Methodology in the Academy," *Organization* 8, no. 2 (2001), 234–42, https://doi.org/10.1177/1350508401082009. For a detailed study of the theory and its application, see Leonie Pihama, "Tihei Mauri Ora: Honouring Our Voices: Mana Wahine as a Kaupapa Māori Theoretical Framework" (PhD diss., University of Auckland, 2001).

43. Nepe, "E hao nei e tneei reanga," 4.

## CHAPTER 1. REEFTON

1. Philippa Mein Smith, *A Concise History of New Zealand* (Cambridge University Press, 2012), 73.

2. Helen Reilly, *Connecting the Country: New Zealand's National Grid* (Steele Roberts, 2008), 11.

3. *Auckland Star*, July 8, 1881, 2.

4. Bernard Rieger, *Technology and the Cultures of Modernity in Britain and Germany 1890–1945* (Cambridge University Press, 2005), 20–22.

5. Alexander Bickerton, "Recent Advances in Electricity," *Proceedings of the New Zealand Institute*, April 28, 1886, 608; "Obituary Notices: Charles Lemon," *Journal of the Institution of Electrical Engineers*, May 30, 1901, 1247.

6. Felicity Barnes, *New Zealand's London: A Colony and Its Metropolis* (University of Auckland Press, 2013), 201–2.

7. "Electric Transmission of Power and Lighting in New Zealand," *Telegraphic Journal and Electrical Review*, February 25, 1887, 184.

8. "Don't Suffer, but Use Dr. Scott's Electric Hairbrush," *Thames Star*, January 17, 1887, 3.

9. *Weekly Press*, December 24, 1886, 1.

10. Nathan Kapoor, "'Who Has Seen the Wind': Imagining Wind Power for the Generation of Electricity in Victorian Britain," *Technology and Culture* 60, no. 2 (2019): 467–93.

11. Rollo Arnold, *New Zealand's Burning: The Settlers' World in the Mid-1880s* (Victoria University Press, 1994), 97.

12. James Belich, *Replenishing the Earth: The Settler Revolution and the Rise of the Anglo-World, 1783–1939* (Oxford University Press, 2009), 12–14.

13. Joanna Boileau, *Chinese Market Gardening in Australia and New Zealand: Gardens of Prosperity* (Springer, 2017), 2–3.

14. Jim McAloon, "Gentlemanly Capitalism and Settler Capitalism: Imperialism, Dependent Development and Colonial Wealth in the South Is-

land of New Zealand," *Australian Economic History Review* 42, no. 2 (2002): 208–9.

15. "Nelson Goldfields," *Otago Witness*, June 15, 1872, 10; "Town and Country," *Lyttleton Times*, June 27, 1872, 2.

16. Len Richardson, *People and Place: The West Coast of New Zealand's South Island in History and Literature* (ANU Press, 2020), 39.

17. "Reefton Share Market," *West Coast Times*, July 17, 1877, 2.

18. David Grant, *Bulls, Bears, and Elephants: A History of the New Zealand Stock Exchange* (Victoria University Press, 1997), 49–50.

19. "News of the Day," *Press*, July 24, 1877, 2.

20. Steven Eldred-Grigg, *Diggers, Hatters, and Whores: The Story of the New Zealand Gold Rushes* (Random House, 2008), 192.

21. "The Inangahua," *The Colonist*, May 14, 1872, 3.

22. John Rosanowski, *Bottled Lightning: The Story of the Reefton Electric Light* (Cornwall & Associates, 2001), 8.

23. "Reefton's Millions," *New Zealand Graphic*, November 22, 1902, 1334.

24. "The Electric Light at Reefton," *Otago Daily Times*, March 9, 1883, 2.

25. John E. Martin, *People, Politics and Power Stations: Electric Power Generation in New Zealand, 1880–1990* (Bridget Williams Books, 1991), 18–20; Reilly, *Connecting the Country*, 17–18.

26. Walter Prince, "Electricity as a Motor," *Inangahua Times*, August 13, 1888, 2.

27. "Electricity and Its Uses," *Grey River Argus*, October 2, 1886, 2.

28. "The Electric Light at Reefton," *Evening Star*, April 24, 1883, 4.

29. "Not a Success," *Timaru Herald*, December 11, 1888, 4.

30. Neil Rennie, *Power to the People: 100 Years of Public Electricity Supply in New Zealand* (Electric Supply Association of New Zealand, 1989), 38–39.

31. Harry Orton, "The History of Underground Power Cables," *IEEE Electrical Insulation Magazine* 29, no. 4 (2013): 52–57.

32. *Inangahua Times*, August 1, 1887, 2.

33. "The Future of Electricity: A Narrow Squeak," *Ashburton Times*, December 4, 1888, 3.

34. "Mining," *Otago Daily Times*, December 14, 1887, 2.

35. Walter Prince, "Electricity as a Motor," *Inangahua Times*, August 13, 1888, 2.

36. "Electric Lighting at Reefton," *Otago Daily Times*, August 6, 1888, 2.

37. *Inangahua Times*, August 1, 1888, 2.

38. "Reefton Electric Lighting," *Thames Star*, August 27, 1888, 4.

39. "Summary for Europe," *New Zealand News*, September 6, 1888, 2; Michael Bryan Schiffer, *Power Struggles: Scientific Authority and Practical Electricity Before Edison* (MIT Press, 2008), 309.

40. E. L. Owen, "Mill Creek Reaches Historic Milestones," *IEEE Power Engineering Review* 17, no. 7 (July 1997): 30–33, https://doi.org/10.1109/MPER.1997.594923; M. Delimar et al., "First Integrated Electric Power System in Croatia," *EUROCON 2007*—The International Conference on "Computer as a Tool," Warsaw, Poland, 2007, 2648–51.

41. *Inangahua Times*, September 10, 1888, 2.

42. "Not a Success," *Timaru Herald*, December 11, 1888, 4.

43. *Inangahua Times*, October 10, 1888, 2.

44. "Reefton Electric Light," *New Zealand Times*, October 15, 1888, 5.

45. "Prince, Walter—Dunedin—Electrical Engineer," Dunedin Bankruptcy Files, R19949228, box 565, folder 252, ANZ.

46. *Inangahua Times*, December 16, 1888, 3.

47. *Inangahua Times*, September 17, 1888, 2.

48. Rosanowski, *Bottled Lightning*, 17.

49. "The Electric Light: Sydney and the Forgotten Wart," *Kumara Times*, November 17, 1888, 3.

50. "Progress in New Zealand," *The Electrician*, November 23, 1888, 68.

51. *The Grey River Argus*, January 22, 1887, 2.

52. Lyman Tower Sargent, "Utopianism and the Creation of New Zealand National Identity," *Utopian Studies* 12, no. 1 (2001): 1–2; Dominic Alessio, "Promoting Paradise: Utopianism and National Identity in New Zealand, 1870–1930," *New Zealand Journal of History* 42, no. 1 (2008): 22–41.

53. Jason Rudy, *Imagined Homelands: British Poetry and the Colonies* (Johns Hopkins University Press, 2017); Daniel P. Graham, "Advertising Empire: Consumerism and the Spatial Imaginary of the British Empire," *Journal of Colonialism and Colonial History* 23, no. 2 (2022), https://doi.org/10.1353/cch.2022.0013.

54. Alan Lester, "Imperial Circuits and Networks: Geographies of the British Empire," *History Compass* 4, no. 1 (2006): 124–41.

55. Keith Robbins, *Great Britain: Identities, Institutions and the Idea of Britishness* (Longman, 1998), 206–7; Simon Featherstone, *Englishness: Twentieth-Century Popular Culture and the Formation of English Identity* (Edinburgh University Press, 2009), 1–8.

56. Michael Adas, *Machines as the Measure of Man* (Cornell University

Press, 1990); David Arnold, *Everyday Technology: Machines and the Making of India's Modernity* (University of Chicago Press, 2013); Clive Dewey, *Steamboats on the Indus: The Limits of Western Technological Superiority in South Asia* (Oxford University Press, 2014); Antoinette Burton, "On the Inadequacy and the Indispensability of the Nation," in *After the Imperial Turn: Thinking with and Through the Nation* (Duke University Press, 2003), 6.

57. Sheila Jasanoff, "Future Imperfect: Science, Technology, and the Imaginations of Modernity," in *Dreamscapes of Modernity: Sociotechnical Imaginaries and the Fabrication of Power*, ed. Sheila Jasanoff and Sang-Hyun Kim (University of Chicago Press, 2015), 4.

58. Liliane Hilaire-Perez and Catherine Verna, "Dissemination of Technical Knowledge in the Middle Ages and the Early Modern Era: New Approaches and Methodological Issues," *Technology and Culture* 47, no. 3 (2006): 543.

59. Charles Hursthouse, *New Zealand or Zealandia, the Britain of the South* (Edward Stanford, 1857).

60. A. W. Reed, *Reed Book of Māori Mythology* (Reed, 2004), 15.

61. Ani Mikaere, "Racism in Contemporary Aotearoa: A Pakeha Problem," in *Colonising Myth, Māori Realities: He Rukuruku Whakaaro* (Huia Publishers, 2011), 92–126.

62. Vincent O'Malley, *The Meeting Place: Māori and Pākehā Encounters: 1642–1840* (Auckland University Press, 2012), 29.

63. A. Grenfell Price, ed., *The Explorations of James Cook in the Pacific: As Told by Selections of His Journals, 1768–1779* (Dover Publications, 1971), 16–17.

64. For a more detailed account of the European encounter between the Māori and Europeans, especially in the nineteenth century, see Vincent O'Malley's *The Meeting Place: Māori and Pakeha Encounters, 1642–1840*; and also, for the later period, see chapters 6 and 7 of Atholl Andersen et al., *Tangata Whenua: A History* (Bridget Williams Books, 2015).

65. Raymond Bunker, "Systematic Colonization and Town Planning in Australia and New Zealand," *Planning Perspectives*, 3, no. 1 (2007): 59–80.

66. Cecilia Morgan, *Building Better Britains: Settler Societies in the British World* (University of Toronto Press, 2016), 37.

67. Patricia Burns and Henry Richardson, eds., *Fatal Success: A History of the New Zealand Company* (Heinemann Reed, 1989).

68. James Belich, *The New Zealand Wars and the Victorian Interpretation of Racial Conflict* (University of Auckland Press, 1986), 312.

69. Phillipa Smith, *Concise History of New Zealand*, 81.

70. Anderson et al., *Tangata Whenua*, 110.

71. Richard Boast and Richard S. Hill, eds., *Raupatu: The Confiscation of Maori Land* (Victoria University Press, 2010).

72. Ernest Dieffenbach, *New Zealand and Its Native Population* (Smith, Elder, 1841), 14; Julie Evans et al., "Australasia: One or Two 'Honorable Cannibals' in the House?" in *Equal Subjects, Unequal Rights: Indigenous People in British Settler Colonies, 1830–1910* (Manchester University Press, 2003), 63–87.

73. George Grey to Lord Stanley, April 22, 1846, cited in Julie Evans, "To Keep Within Proper Bounds . . .": Edward Eyre and the Colonised Peoples of Australia New Zealand, and the Caribbean," PhD diss., University of Melbourne, 1998, 174.

74. Nicola Rowan Wheeler and Janine Hayward, eds., *Treaty of Waitangi Settlements* (Bridget Williams Books, 2012).

75. Smith, *Concise History of New Zealand*, 82.

76. Peter Holland et al., "Pioneering Grassland Farming: Pragmatism, Innovation, and Experimentation," in *Seeds of Empire: The Environmental Transformation of New Zealand*, ed. Tom Brooking and Eric Pawson (I. B. Tauris, 2011), 51–52.

77. Richard Boast, *Buying and Selling the Land: Government and the Māori Land in the North Island, 1865–1921* (Victoria University Press, 2008), 19–22.

78. George Butler Earp, *The Emigrant's Guide to New Zealand: Comprising Every Requisite Information for Intending Emigrants*, 5th ed. (W. S. Orr and Co., 1849), 48.

79. Lucy Sargisson and Lyman Tower Sargent, *Living Utopias: New Zealand's Intentional Communities* (Ashgate, 2004), 1–37.

80. D. George Boyce, "Gladstone and Ireland," in *Gladstone*, ed. Peter John Jagger (Hambledon Press, 1998), 118–21.

81. William E. Gladstone, Speech of the Rt. Hon. William E. Gladstone, M.P. on the Second Reading of the New Zealand Constitution (John W. Parker and Son, 1852), 18.

82. George Grey, "Colonial Dispatch," August 31, 1846.

83. Barnes, *New Zealand's London*, 65.

84. Harold Passer, *The Electrical Manufacturers, 1875–1900: A Study in Competition, Technical Change, and Economic Growth* (Harvard University Press, 1953), 110.

85. Frederick Young, ed., *Imperial Federation of Great Britain and Her Colonies* (S. W. Silver and Co., 1876), 157.

86. Julius Vogel, "New Zealand and the South Sea Islands, and Their Relation to the Empire," *Proceedings of the Royal Colonial Institute* 9 (1878): 42.

87. Hursthouse, *New Zealand or Zealandia*, 5.

88. For interested readers, he declared, "I should have much pleasure in attempting to afford it by replying to any communication sent to my town address, 28, Thavies Inn, London." Hursthouse, *New Zealand or Zealandia*, 8.

89. James Belich, "Colonization and History in New Zealand," *The Oxford History of the British Empire*, vol. 5, ed. Robin W. Winks (Oxford University Press, 2001), 182.

90. Hursthouse, *New Zealand or Zealandia*, 144.

91. Hursthouse, *New Zealand or Zealandia*, 145.

92. Thomas Wayth Gudgeon, *The Defenders of New Zealand: Being a Short Biography of Colonists Who Distinguished Themselves in Upholding Her Majesty's Supremacy* (H. Brett, 1887), 26.

93. "New Zealand Institute," *Wellington Independent*, September 7, 1869, 3.

94. Felicity Barnes, *New Zealand's London*, 81.

95. "The Electric Light at Reefton," *Otago Daily Times*, March 9, 1883, 2.

96. William Thomson, "Address to the Mathematical and Physical Science Section of the British Association," *The Chemical News and Physical Sciences Journal* 44, no. 138 (1881), 135–37; Silvanus P. Thompson, "Storage of Electricity," *The Electrician* 8, no. 2 (1881): 22–55; C. W. Siemens, "Presidential Address," *Telegraphic Journal and Electrical Review* 11 (1882): 146.

97. For Jevons and many others, coal was the center of British industrial capability. With the expansion of American and other European industrial capabilities, Jevons warned, the British might no longer be able to keep up given the coal resources available to Great Britain. William Stanley Jevons, *The Coal Question: An Inquiry Concerning the Progress of the Nation* (MacMillan and Co., 1865), 105. On the other hand, John Ruskin expressed concern about the effect industrialization had on British citizens, especially the neglect of the lower classes, arguing that such disenfranchisement would destroy the empire from within. John Ruskin, *The Works of John Ruskin*, vol. VI, *The Crown of Wild Olive* (George Allen, 1882), 150–53.

98. "Home and Colonial Topics," *New Zealand Times*, July 31, 1882, 3.

99. "Water Power," *Lyttleton Times*, February 14, 1880, 5.

100. "Mr. Prince's Lecture," *Inangahua Times*, December 1, 1886, 2.

101. Kapoor, "'Who Has Seen the Wind.'"

102. Crosbie Smith and M. Norton Wise, *Science and Empire: A Biographical Study of Lord Kelvin* (Cambridge University Press, 1989), 799.

103. Reefton Tourism, "Reefton Homepage," accessed June 12, 2023, https://www.reefton.co.nz/discover/our-history.

104. "Reefton in Summer," Chorus, March 15, 2023, last accessed May 28, 2026, https://www.chorus.co.nz/cabinet-art-project/reefton-summer.

105. As of 2018, 123 Māori live in Reefton. "New Zealand 2018 Census," accessed January 3, 2023, https://www.stats.govt.nz/tools/2018-census-place-summaries/reefton.

106. "Land Development on West Coast," *Press*, March 31, 1956, 7.

107. Dipesh Chakrabarty, *Provincializing Europe: Postcolonial Thought and Historical Difference—New Edition* (Princeton University Press, 2009), 45–46; On Barak, *On Time: Technology and Temporality* (2013), 36.

## CHAPTER 2. MINING FOR POWER

1. Lloyd Carpenter, "A Rich Myth, Gold, and Narrative: Aspects of the Central Otago Gold Rush, 1862–2012" (PhD diss., Canterbury University, 2013), 13.

2. *Appendix to the Journal of the New Zealand House of Representatives*, 1866 D14: 4.

3. Report by Mr. Mining Surveyor Wright, to the Provincial Government of Otago, *AJHR*, 1866 Session I, D-14, 4.

4. J. C. F. Johnson, *A Practical Treatise for Prospectors, Miners, and Students* (Charles Griffin, 1896), 28–30.

5. Stephen Tuffnell, "Engineering Inter-Imperialism: American Miners and the Transformation of Global Mining, 1871–1910," *Journal of Global History* 10, no. 1 (2015): 53.

6. *Wright's Australian and American Commercial Directory and Gazetteer* (George Wright, 1881), 397.

7. Peter Petchey, *Bullendale: The Phoenix Company's Hydro-Electric Power System Site E40/28, Powerhouse Reconstruction* (Southern Archeology Ltd., 2022), 13.

8. "George H. F. Ulrich," Trove Database, National Library of Australia, accessed March 27, 2022, https://trove.nla.gov.au/people/1238621?c=people.

9. "Upper Shotover," *Evening Star*, July 16, 1874, 2 (Supplement).

10. "The Mines," *Otago Witness*, September 25, 1880, 9; "Mining Report," *The Colonist*, February 18, 1884, 3; "Mining Times," *Tuapeka Times*, October 29, 1884, 5.

11. "Quartz Mining at Skipper's," *Cromwell Argus*, October 31, 1871, 6.

12. "The Skipper's Reefs and Their History," *Lake Wakatip Mail*, December 3, 1884, 5.

13. *Appendix to the Journal of the New Zealand House of Representatives* (1886) C-4: 19.

14. "Untitled," *Otago Witness*, October 10, 1885, 8; Miles Pierce, "Early Electricity Supply in Melbourne," *Third Australasian Engineering Heritage Conference IPENZ* (2009): 1–10.

15. *Appendix to the Journal of the New Zealand House of Representatives* (1901) C-1: 125.

16. Peter Petchey, *Bullendale: The Phoenix Company's Hydro-Electric Power System Site E40/28, Powerhouse Reconstruction* (Southern Archeology Ltd., 2022), 15.

17. Pierre-Louis Viollet, "From the Water Wheel to Turbines and Hydroelectricity: Technological Evolution and Revolutions," *Comptes Rendus-Mécanique* 345, no. 8 (2017): 576.

18. Peter Petchey, *Bullendale: The Phoenix Company's Hydro-Electric Power System Site E40/28, Powerhouse Reconstruction* (Southern Archeology Ltd., 2022), 15–16.

19. There is some discrepancy here about the number of stampers at this mine powered by electricity. At one point, there were thirty stamps, but the electric motor only ran twenty of them at first. *Appendix to the Journal of the New Zealand House of Representatives* (1886) C4: 19.

20. "Wakatipu Mining Notes," *Otago Witness*, January 23, 1886, 18.

21. *Appendix to the Journal of the New Zealand House of Representatives* (1887) C5: 46.

22. Earlier articles of praise resemble the one on page 2 of the *Grey River Argus*, April 21, 1875; the quote is from "Upper Shotover: The Success of the Electric Machinery at the Phoenix Mine," *Lake Wakatip Mail*, February 5, 1886, 2.

23. "Mining Intelligence," *Lake Wakatip Mail*, April 30, 1886, 5.

24. "Transmission of Power by Electricity," *Otago Witness*, January 9, 1886, 11.

25. "Mining Enterprise in Otago," *Inangahua Times*, August 13, 1886, 2.

26. *Appendix to the Journal of the New Zealand House of Representatives* (1887) C5: 47; George J. Binns, "On Lighting in New Zealand Mines," *Transactions of the North of England Institute of Mining and Mechanical Engineers* 41 (1893): 123–25.

27. Silvanus Thompson, *Dynamo-Electric Machinery: A Manual for*

*Student of Electrotechnics* (E.& F. N. Spon., 1886) 113–15. Peter Petchey, "Industrial Archeology of Power: The Gold Mining Industry's Search for Power and the Development of Early Hydro-Electric Generation in New Zealand," in *Finding Our Recent Past: Historical Archeology in New Zealand*, ed. M. Campbell et al. (New Zealand Archeological Association, 2014), 207.

28. Helen Reilly, *Connecting the Country: New Zealand's National Grid, 1886–2007* (Steele Roberts, 2008), 232–35; "The Impact of Our Climate on Hydro Generation," Electricity Authority, February 27, 2023, 18–19, https://www.ea.govt.nz/news/eye-on-electricity/the-impact-of-our-climate-on-hydro-generation/; Peter Petchey, "New Zealand's Technological Participation in the International Goldfields: The Example of the Stamper Battery," *Technology and Culture* 60, no. 2 (2019): 494–522.

29. *Gold Mines and Mining in California: A New Gold Era Dawning on the State* (George Spaulding and Co., 1885); "Mysore Gold Mine," *The Mining Journal*, January 7, 1888, 20; Frederick H. Hatch and J. A. Chalmers, *The Gold Mines of the Rand* (Macmillan and Co., 1895).

30. Jim McAloon, "Resource Frontiers, Environment, and Settler Capitalism," in *Making a New Land: Environmental Histories of New Zealand*, ed. Eric Pawson and Tom Brooking (University of Otago Press, 2013), 52–53.

31. Tamara S. Wagner, ed., *Domestic Fiction in Colonial Australia and New Zealand* (Pickering and Chatto, 2014), 6–11.

32. "A Model Colony," *The New Zealand Journal*, October 24, 1840, 250; 192.

33. Edward Gibbon Wakefield, "Report of the Select Committee on South Australia," *Parliamentary Papers Reports from Committees* (George Didsbury Press, 1841), 332–39.

34. E. A. Wrigley and R. S. Schofield, *The Population History of England 1541–1871*, 2nd ed. (Arnold, 1981), 208–9.

35. Charles Feinstein, "Pessimism Perpetuated: Real Wages and the Standard of Living in Britain Before and After the Industrial Revolution," *Journal of Economic History* 58, no. 2 (1998): 625–58.

36. "Arthur Clayden," *Te Ara: The New Zealand Encyclopedia*, last accessed March 28, 2020, https://teara.govt.nz/en/biographies/2c19/clayden-arthur.

37. Charles Hursthouse, *New Zealand or Zealandia: The Britain of the South* (Edward Stanford, 1957), 144.

38. Arthur Clayden, *The England of the Pacific or as an English Middle-Class Emigration-Field* (Wyman and Sons, 1879), 55.

39. Ewout Frankema, "Raising Revenue in the British Empire, 1870–1940," *Journal of Global History* 5, no. 3 (2010): 469.

40. L. Gillbank, "The Origins of the Acclimatisation Society of Victoria: Practical Science in the Wake of the Gold Rush," *Historical Records of Australian Science* 6, no. 3 (1986): 359–74.

41. "The Voyage of the 'Novara,'" *The Spectator*, September 28, 1861, 21.

42. Ferdinand Ritter von Hochstetter, *New Zealand: Its Physical Geography, Geology, and Natural History* (J. G. Cotta, 1867), 25.

43. Stevan Eldred-Grigg, *Diggers, Hatters, and Whores: The Story of the New Zealand Gold Rushes* (Random House New Zealand, 2011); Jock Phillips, History of Immigration—Miners," Te Ara: The Encyclopedia of New Zealand, accessed May 10, 2023, http://www.TeAra.govt.nz/en/history-of-immigration/page-7.

44. Alfred George Warnford Lock, *Gold: Its Occurrence and Extraction* (London: E. &F. N. Spon., 1882), 517.

45. Peter Petchey, *Gold and Electricity: Archeological Survey of Bullendale, Otago* (New Zealand Department of Conservation, 2006), 5.

46. "Mining Intelligence," *Lake Wakatip Mail*, October 15, 1886, 5.

47. "Transmission of Power by Electricity," *Otago Witness*, January 9, 1886, 11.

48. C. W. Siemens, "Presidential Address," *Telegraphic Journal and Electrical Review* 11 (1882): 146; C. Scott Snell, "The Sea—A Source of Power," *Telegraphic Journal and Electrical Review* 11 (1882): 454–55; Edward Hull, *Our Coal Resources at the Close of the Nineteenth Century* (E. & F. N. Spon., 1897), 136–38; Stephen Gray, *Steam Power and Sea Power: Coal, The Royal Navy, and the British Empire, c. 1870–1914* (Palgrave, 2018), 20; Nathan Kapoor, "Who Has Seen the Wind," *Technology and Culture* 60, no. 2 (2019): 467–93.

49. Marian Kent, *Oil and Empire: British Policy and Mesopotamian Oil 1900–1920* (MacMillan, 1976), 1–2; On Barak, "Outsourcing: Energy and Empire in the Age of Coal, 1820–1911," *International Journal of Middle East Studies* 47 (2015): 425–45.

50. *Appendix to the Journal of the New Zealand House of Representatives* (1885) C2: 11.

51. Much of the telegraph industry in the South Island was based in Dunedin and run by the Colonial Telegraph Company. "Telegraphy in New Zealand," *Proceedings of the Institution of Electrical Engineers* 9 (1880): 29–30.

52. Rob Aspden, "Electricity Supply in New Zealand," *Proceedings of the Electrical Engineering Advisory Committee* 14, no. 2 (1988): 187.

53. J. G. Black, "Goldfields Classes and Schools of Mines," *Report on the Mining Industry of New Zealand* (George Didsbury, 1887), 269–70.

54. Graeme Gooday and Morris Low, "Technology Transfer and Cultural Exchange: Western Scientists and Engineers Encounter Late Tokugawa and Meiji Japan," *Osiris* 13 (1998): 99–128; Ruth Barton, "'Men of Science': Language, Identity, and Professionalization in the Mid-Victorian Scientific Community," *History of Science* 41 (2003): 73–119; Michael Bryan Schiffer, "The Electrical Lighthouse in the Nineteenth Century: Aid to Navigation and Political Technology," *Technology and Culture* 46, no. 2 (2005): 305.

55. Harry James Eisenman, "Charles F. Brush: Pioneer in Electrical Technology" (PhD diss., Case Institute of Technology, 1967), 110.

56. "R. E. Fletcher Obituary," *Journal of the Institute of Electrical Engineers*, 1936, 692–93; *Grey River Argus*, October 2, 1886, 2.

57. "Transmission of Electric Power for Gold Mining," *Thames Adviser*, June 14, 1886, 2; Joseph W. Buell, "The Coming Motive Power: Is Compressed Air to Become the Rival of Electricity?" *California Journal of Mines and Geology* 12–13 (1893): 721–25; "South Africa Again," *Engineering and Mining Journal* 60 (December 7, 1895): 535–36.

58. *The Handbook of New Zealand Mines* (George Didsbury, 1887), 48.

59. "Electrical Transmission of Power in New Zealand," *Telegraphic Journal and Electrical Review* 19, August 6, 1886, 134.

60. "Electrical Transmission of Power in New Zealand," 135.

61. Timothy Cooper, "Peter Lund Simmonds and the Political Economy of Waste Utilization in Victorian Britain," *Technology and Culture* 52, no. 1 (2011): 21–44.

62. "Mining Intelligence," *The Star*, August 24, 1861, 2.

63. Lloyd Carpenter, "A Rich Myth, Gold, and Narrative: Aspects of the Central Otago Gold Rush, 1862–2012" (PhD thesis, Canterbury University, 2013). 44–71.

64. Jill Hamel, *The Archeology of Otago* (Department of Conservation, 2001), 184–87; Keir Reeves, "Sojourners or a New Diaspora: Economic Implications of Chinese Miners to the South-Pacific Goldfields," *Australian Economic History Review* 50, no. 2 (2010): 178–92. In particular, one Otago mining town, Arrowtown, to the southeast of Bullendale, had a small Chinese settlement. Archaeologists have found Chinese Daoist and Buddhist temples throughout Otago, most notably in Arrowtown. James Ng, "The Sojourner Experience: The Cantonese Goldseekers in New Zealand, 1865–1901," in *Unfolding History, Evolving Identity: The Chinese in New Zealand*, ed. Manying Ip (Auckland University Press, 2003), 14; "Arrowtown Chi-

nese Settlement," Department of Conservation, accessed March 15, 2018, http://www.doc.govt.nz/Documents/about-doc/concessions-and-permits/conservation-revealed/arrowtown-chinese-settlement-lowres.pdf.

65. John C. Weaver, "Land Seizures and Aftershocks: 'A True Lesson in History' on Anglo-Colonialism," in *Raupatu: The Confiscation of Māori Land*, ed. Richard Boast and Richard Hill (Victoria University Press, 2009), 51.

66. A. Wanhalla and L. Paterson, "'Tangled Up': Intimacy, Emotion, and Dispossession in Colonial New Zealand," in *Intimacies of Violence in the Settler Colony*, ed. P. Edmonds and A. Nettelbeck (Palgrave Macmillan, 2018).

67. Vincent O'Malley, *The Meeting Place: Māori and Pākehā Encounters, 1642–1840* (Auckland University Press, 2012), 110–11.

68. Atholl Anderson, "Old Ways and New Means, 1810–1830," in *Tangata Whenua: A History* (Bridget Williams Books, 2015), 228–29.

69. O'Malley, *Meeting Place*, 131.

70. Atholl Anderson et al., eds., *Tangata Whenua: A History* (Bridget Williams Books, 2015), 173.

71. Richard Boast and Richard Hill, eds., *Raupatu: The Confiscation of Māori Land* (Victoria University Press, 2009), 5–6.

72. F. D. Fenton, "Observations on the State of the Aboriginal Māori Inhabitants of New Zealand," *Journals of the Statistical Society* (1860): 508–41; James Belich, *Making Peoples: A History of New Zealanders from Polynesian Settlement to the End of the Nineteenth Century* (University of Hawaii Press, 1996), 464.

73. Patricia Burns, *Fatal Success: A History of the New Zealand Company* (Heinemann Reed, 1989).

74. Robert J. Miller et al., *Discovering Indigenous Lands: The Doctrine of Discovery in the English Colonies* (Oxford University Press, 2010), 212–13.

75. "Population and Houses," Census of the Colony of New Zealand presented to the Colonial Secretary, March 28, 1886, Statistics New Zealand, accessed February 1, 2019, https://www3.stats.govt.nz/Historic_Publications/1886-census/Results-of-Census-1886/1886-results-census.html?_ga=2.86348508.511668155.1552436752-659471187.1552436752#d50e322.

76. Robert Pemberton, *The Happy Colony: Dedicated to the Workmen of Great Britain* (Saunders and Otley, 1854), 48. Here I borrow from Ling Zhang and Xiangli Ding's work on the Yellow River in the twelfth and twentieth centuries, respectively, in which the intertwining of the environment and state hinged on consumption of water in problematic ways. Ling Zhang,

*The River, The Plain, and the State: An Environmental Drama in Northern Song China, 1048–1128* (Cambridge University Press, 2016), 181; Xiangli Ding, "'The Yellow River Comes from Our Hands': Silt, Hydroelectricity, and the Sanmenxia Dam, 1929–1973," *Environment and History* 27, no. 4 (2021): 668.

77. "Electric Power in Mining," *The Press*, July 12, 1888, 3.

78. Helen Reilly, *Connecting the Country: New Zealand's National Grid, 1886–2007* (Steele Roberts, 2008), 232–35; "The Impact of Our Climate on Hydro Generation," Electricity Authority, February 27, 2023, last accessed July 18, 2023, https://www.ea.govt.nz/news/eye-on-electricity/the-impact-of-our-climate-on-hydro-generation/.

79. Ministry of Business, Innovation, and Employment, "Data Tables for Electricity," last accessed June 21, 2023, https://www.mbie.govt.nz/building-and-energy/energy-and-natural-resources/energy-statistics-and-modelling/energy-statistics/electricity-statistics/.

## CHAPTER 3. THE STATION AT WORLD'S END

1. Telegrams sent to Prime Minister Joseph Ward, "Congratulations," *Bush Advocate* 19, no. 831, September 27, 1907, 5.

2. David. E. Nye, "Electrifying Exhibitions: 1880–1939," in *Fair Representations: World Fairs and the Modern World*, ed. Robert Rydell and Nancy Gwinn (Free University Press, 1994), 140–56.

3. Graeme Gooday, *Domesticating Electricity: Technology, Uncertainty, and Gender, 1880–1914* (Pickering and Chatto, 2008): 121–22.

4. Kenneth O. Morgan, ed., "Chronology," in *The Oxford History of Britain* (Oxford University Press, 2010), 760–65.

5. Samuel Leonard Tilley (1818–1896) of New Brunswick proposed the title of Dominion as part of the British North America Act (1867). William Parr Greswell, *History of the Dominion of Canada* (Clarendon Press, 1890), 212.

6. G. A. Wood, "The Former Dominion of New Zealand," *Political Science* 26, no. 1 (1974): 2–10.

7. The 1901 census divided the population into Māori (43,112) and non-Māori (772,719), including Europeans, Chinese, and others. "A Census of the Colony of New Zealand," March 31, 1901; James Belich, *Paradise Reforged: A History of the New Zealanders from the 1880s to the Year 2000* (University of Hawaii Press, 2001), 16–17.

8. Edward Gibbon Wakefield, *Twenty-Third Report of the Court of Directors of the New Zealand Company* (New Zealand House, 1847), 2; William

David McIntyre and W. J. Gardener, eds., *Speeches and Documents on New Zealand's History* (Clarendon Press, 1971), 21.

9. Michel Foucault, *The Birth of Biopolitics: Lectures at the Collége de France, 1978–1979* (Picador, 2010), 70; Brian Larkin, "The Politic and Poetics of Infrastructure," *Annual Review of Anthropology* 42 (2013): 328.

10. "Lord Stanley's Speech," *New Zealand Colonist and Port Nicholson Advertiser*, August 9, 1842, 4; "Untitled," *The Nelson Examiner*, September 10, 1842, 106.

11. "Governor T. Gore Browne to Sir George Grey," March 12, 1856, published in *Accounts and Papers of the House of Commons* 46 (George Edward Eyre and Spottiswoode, 1860), 193.

12. Charles Hursthouse, "Double Government," in *Letters on New Zealand Subjects* (Edward Stanford, 1865), 63–64.

13. "Letter to the Editor," *New Zealander*, May 8, 1861, 3; C. W. Burnell, "The Māori and the Moa," *Ashburton Herald*, March 5, 1880, 5; "City Māori Problem," *Gisborne Herald*, April 13, 1950, 4.

14. Alan Ward, *An Unsettled History: Treaty Claims in New Zealand Today* (Bridget Williams Books, 2015), 162.

15. Harold E. Raugh Jr., ed., *The Victorians at War, 1815–1914: An Encyclopedia of British Military History* (ABC-Clio, 2004), 82.

16. Morgan, *Oxford History of Britain*, 561–63.

17. James Belich, *The Victorian Interpretation of Racial Conflict: The Maori, the British, and the New Zealand Wars* (McGill-Queen's University Press. 1989), 23–24; Richard Hill, *State Authority, Indigenous Autonomy: Crown-Māori Relations in New Zealand Aotearoa, 1900–1950* (Victoria University Press, 2004), 37–42.

18. Henry Sewell and David McIntyre, eds., *The Journal of Henry Sewell*, vol. 2 (Whitcoulls Publishers, 1980), 320.

19. "Home News," *Pall Mall Budget*, November 26, 1869, 30–32; "News of the Week," *The Spectator*, December 4, 1869, 1414–15.

20. William Gladstone, "England's Mission," *The Nineteenth Century*, August 25, 1878, 560–85; C. C. Eldridge, *England's Mission: The Imperial Idea in the Age of Gladstone and Disraeli, 1868–1880* (Macmillan, 1973), 98–119.

21. Eqout Frankema, "Raising Revenue in the British Empire, 1870–1940: How 'Extractive' Were Colonial Taxes?," *Journal of Global History* 5, no. 3 (2010): 447–77; Felicity Barnes, *New Zealand's London: A Colony and Its Metropolis* (Auckland University Press, 2012), 17.

22. Barnes, *New Zealand's London*, 9.

23. Julius Vogel, *Appendix to the House of Representatives* (George Didsbury, 1874), A-3, 9–10.

24. "Confederation and Annexation Act," *New Zealand Herald*, October 8, 1883, 11 (Supplement).

25. Raewyn Dalziel, "Southern Islands: New Zealand and Polynesia," in *The Oxford History of the British Empire: The Nineteenth Century*, ed. Andrew Porter (Oxford University Press, 1999), 591.

26. M. D. Burgess, "Lord Rosebery and the Imperial Federation League, 1884–1893," *New Zealand Journal of History* 13 (1979): 165–81.

27. "Dominion of New Zealand," *The Feilding Star*, September 26, 1907, 2.

28. "Dominion Day Celebrations in New Zealand: The Proclamation, Festivities in the City, Sights, and Sounds, Message from the King," *Evening Post*, September 26, 1907, 7.

29. "Property Assessment Bill," *Parliamentary Debates* (G. Didsbury, 1889), 239–40.

30. James Belich, *Making Peoples* (University of Hawaii Press, 2001), 437.

31. Peter Holland et al., "Remaking the Grasslands of the Open Country," in *Environmental Histories of New Zealand*, ed. Eric Pawson and Tom Brooking (Oxford University Press, 2002), 69.

32. "Immigration and Public Works Act," September 12, 1870.

33. Suction gas (or producer gas) systems are those that generated combustible gas from solid fuels like coal, coke, or biomass. The chamber in which this happens creates a vacuum that literally "sucks" the air and steam through the hot fuel. "The Suction Gas Producer," *The North Western Advocate and the Emu Bay*, July 18, 1908, 2; Helen Reilly, *Connecting the Country: New Zealand's National Grid 1886–2007* (Steele Roberts, 2008), 24–25.

34. "Electric Light," *Star*, December 1, 1888, 3. The Stratford Electric Company formed in May 1898 to carry out the town's electrification by water power. "The Electric Light Proposals," *Northern Advocate*, May 21, 1898, 3. After the success of Stratford's electrification, the trustees of Parihaka Pa decided to install public electric lighting. John Martin, *People, Politics, and Power Stations* (Electricity Corporation of New Zealand, 1991): 31–32.

35. "Electric Motive Power Bill," *Parliamentary Debates* 96 (1896): 237–38.

36. "Electric Motive Power Act," October 17, 1896, 60, no. 47, 149.

37. "Electric Lighting in the Colonies," *New Zealand Herald*, June 13,

1885, 1 (Supplement); Rosemary Britten, *Lake Coleridge: The Power, the People, the Land* (Hazard Press, 2000): 100–101.

38. "Untitled," *Cromwell Argus*, June 29, 1886, 2; "New Zealand Industrial Exhibition," *New Zealand Times*, August 14, 1889, 5.

39. "Artesian Motor Power," *Star*, July 1, 1891, 1; "The Plague," *Thames Star*, May 15, 1900, 2.

40. "The Gas Question in Christchurch," *South Canterbury Times*, July 17, 1888, 2; "Mr. J. L. Scott's Idea," *Star*, May 29, 1899.

41. "The New City Surveyor," *The Press*, December 12, 1900, 2.

42. D. M. Calder, "Plant Ecology of Subalpine Shingle River-Beds in Canterbury, New Zealand," *Journal of Ecology* 49, no. 3 (1961): 581–94.

43. "A Big Scheme," *Lyttleton Times*, June 11, 1900, 3.

44. T. E. Taylor, "Harness the Waimakariri," *The Press*, October 27, 1900, 3.

45. Taylor, "Harness the Waimakariri."

46. Taylor, "Harness the Waimakariri."

47. Renfrew Christie, *Electricity, Industry, and Class in South Africa* (State University of New York Press, 1984), 5; Hughes, *Networks of Power*, 262–64; Rennie, *Power to the People: 100 Years of Public Electricity Supply in New Zealand* (Electricity Supply Association of New Zealand, 1989), 233–34.

48. "Electricity as a Motive Power," *Timaru Herald*, October 23, 1900, 4.

49. "Town and Country," *Lyttleton Times*, January 18, 1904, 4.

50. "The Waimakariri Scheme," *Lyttleton Times*, May 14, 1901, 2; "Loss to Dunedin," *New Zealand Herald*, November 29, 1928, 8.

51. "County Merger?" *The Press*, June 11, 1927, 14.

52. The letter is signed "progressive." This likely means it was written by Taylor. "Waimakariri Power Scheme," *Lyttleton Times*, August 1901, 7.

53. Serge Paquier, "Turrettini, Théodore," in *Dictionnaire Historique de la Suisse*.

54. "City of Christchurch Electric Power and Loan Empowering Bill," in *Parliamentary Debates: House of Representatives* (John Mackay, 1902), 250.

55. "Christchurch City Council," *Lyttleton Times*, November 22, 1902, 9.

56. "Harnessing the Waimakariri," *Lyttleton Times*, August 29, 1902, 5; "Interview with Mr. T. E. Taylor," *The Press*, September 18, 1902, 5; "River Eyre Protective Works," *The Press*, November 13, 1902, 5.

57. *Water-Power Act 1903* (New Zealand) 3 EDW VII 1903 No. 26, 54–55, http://www.nzlii.org/nz/legis/hist_act/wa19033ev1903n26248/.

58. "New Zealand Water Powers, Etc.," *Appendix to the Journal of the House of Representatives*, D-7 1901 Session 1, 2.

59. "Mr. Hancock's Report on Water Power," *New Zealand Herald*, November 1, 1904, 5.

60. "Electrical Energy: New Zealand's Wonderful Possibilities," *The Press*, October 31, 1904, 10.

61. "Christchurch Monopolies," *Lyttleton Times*, May 19, 1905, 3.

62. "Electrical Transmission of Power," *Evening Post*, October 31, 1904, 2.

63. Bill Pierre, *The North Island Main Trunk* (Reed, 1981), 289–90.

64. Belich, *Paradise Reforged*, 61.

65. "Power for Christchurch," *Star*, January 13, 1910, 1.

66. "Water Power," *Lyttleton Times*, February 4, 1909, 9.

67. "State Water Power," *Lyttleton Times*, August 24, 1910, 7.

68. John E. Martin, "Birks, Lawrence," *Te Ara: Encyclopedia of New Zealand*, accessed June 11, 2018, https://teara.govt.nz/en/biographies/3b34/birks-lawrence.

69. "The Lake Coleridge Hydroelectric Scheme," *The Press*, December 3, 1910, 10.

70. Frederick Black, "Water Power Scheme: Government Proposals Criticized," *Star*, October 18, 1910, 4, https://paperspast.natlib.govt.nz/newspapers/TS19101018.2.62.

71. "Aid to Water-Power Works Bill," *Parliamentary Debates*, November 1, 1910, 191–99; Aid to Water Power Works Act," November 21, 1910.

72. "Water Power," *Bush Advocate*, November 19, 1910, 4.

73. After the completion of the system, several farmers formed the Murchison Trusteeship to demand compensation from the land lost to construction and power generation. They eventually received £1,474. Mountain Lily [pseud.], "Letter the Editor: Lake Coleridge Township," *The Press*, June 1, 1912, 14; "Compensation Claim: Land at Lake Coleridge," *Lyttleton Times*, October 8, 1915, 4.

74. W. Wilson, "Hydro-Electric Power in New Zealand," *Journal of the Institute of Electrical Engineers*, January 7, 1916, 283; Reilly, *Connecting the Country*, 38.

75. Martin, *People, Politics, and Power Stations*, 48.

76. "Strike at Lake Coleridge," *West Coast Times*, September 24, 1912, 3.

77. "Glaciation at Lake Coleridge," *Star*, September 4, 1913, 1; R. Speight, "Ice Wasting and Glacier Retreat," *Journal of Geomorphology* 3 (1940): 131–41.

78. "Miners Complain," *New Zealand Times*, July 12, 1912, 8.

79. "Miners Complain," *New Zealand Times*, July 12, 1912, 8.

80. "Lake Coleridge Notes," *Press*, February 21, 1913, 9.

81. "Town and Country," *Timaru Herald*, January 17, 1913, 6; "State Electric Power," *Wanganui Herald*, March 31, 1913; "Notices," *Greymouth Evening Star*, April 20, 1914, 4.

82. "The City's Electricity, Progress of the Great Scheme from the Lake to the Citizen's Home," *Lyttleton Times*, July 14, 1914, 15.

83. "Cheap Current: The Lake Coleridge Power," *Press*, July 10, 1913, 5.

84. "Lake Coleridge Explored," *Sun*, April 4, 1914, 14; Peter Frank and Jim McAloon, *Labour: The New Zealand Labour Party, 1916–2016* (Victoria University Press, 2016).

85. "Lake Coleridge Current," *Press*, October 21, 1914, 2.

86. "The White Coal: A Visit to Lake Coleridge," *Manawatu Times*, November 29, 1917, 6.

87. "Lake Coleridge. The Supply of Power: Is the Government a Competitor?" *Lyttleton Times*, March 5, 1914, 3.

88. William Fraser, "Public Works Statement," *Appendix to the Journals of the House of Representatives*, 1915, Session I, D-01, xi–xii.

89. "Lake Coleridge," *Lyttleton Times*, November 26, 1914, 6.

90. "City's Electricity," *Star*, November 26, 1914, 8.

91. "Lake Coleridge," *Lyttleton Times*, November 26, 1914, 6; Thomas Hughes, *Networks of Power: Electrification in Western Society, 1880–1930* (Johns Hopkins University Press, 1983), 292. During the nineteenth and early twentieth centuries, hydropower advocates frequently referred to waterways with the potential of hydroelectric generation as "white coal"—not to be confused with the white coal that refers to charcoal made by the drying of wood.

92. Bruce Railsback, "Rain, Riches, and Empire: The Relationship Between Nations Ruling Distant Lands, Nations of Great Wealth, and Regions of Regular Moderate Atmospheric Precipitation," *Weather, Climate, and Society* 9, no. 3 (2017): 455–69.

93. Sara Pritchard, "From Hydroimperialism to Hydrocapitalism: 'French' Hydraulics in France, North Africa, and Beyond," *Social Studies of Science* 42, no. 4 (2012): 591–615.

94. J. M. D. Meiklejohn, *The British Empire: Its Geography, Resources, Commerce, Land-ways, and Water-ways*, 6th ed. (Alfred M. Holden, 1899), 16.

95. John Broich, "Engineering the Empire: British Water Supply Systems and Colonial Societies, 1850–1900," *Journal of British Studies* 46 (2007): 346–65.

96. Dorotea Gucciardo, "The Powered Generation: Canadians, Electricity, and Everyday Life" (PhD diss., University of Western Ontario, 2011), 49–50.

97. Daniel McFarlane and Peter Kitay, "Hydraulic Imperialism: Hydroelectric Development and the Treaty 9 in the Abititi Region," *American Review of Canadian Studies* 46, no. 3 (2016): 380–97; Jane Griffith, "Do Some Work for Me: Settler Colonialism, Professional Communication, and Representation of Indigenous Water," *Decolonization: Indigeneity, Education, and Society* 7, no. 1 (2018): 132–57.

98. By framing it this way, I am building on Patrick Wolfe's seminal work and Erika Marie Bsumek's recent study of the Glen Canyon Dam. Wolfe, *Settler Colonialism and the Transformation of Anthropology: The Politics and Poetics of an Ethnographic Event* (Cassell, 1999); Bsumek, *The Foundations of Glen Canyon Dam: Infrastructures of Dispossession on the Colorado Plateau* (University of Texas Press, 2023), 5.

99. D. Williams, "Ko Aotearoa Tenei: Law and Policy Affecting Māori Culture and Identity," *International Journal of Cultural Property*, 20 (2013): 311–31.

100. "Election Campaign: Premier in Christchurch," *Grey River Argus*, November 26, 1914, 7.

101. "About Us," TrustPower, accessed November 23, 2023, https://web.archive.org/web/20220127185844/https://www.trustpower.co.nz/Getting-To-Know-Us/Our-History.

102. Rennie, *Power to the People*; Reilly, *Connecting the Country*.

103. Reilly, *Connecting the Country*, 25.

104. F. Edward and M. Hård, "Maintaining the Local Empire: The Public Works Department in Dar es Salaam, 1920–60," *Journal of Transport History* 40, no. 1 (2019): 43.

## CHAPTER 4. "THE MEGAWATTASAURUS"

*Epigraph*: Robert Te Kotahi Mahuta, *Planning Issues and the Huntly Power Station* (Center for Māori Studies and Research, University of Waikato, 1978), 11.

1. "Local and General," *Sun*, May 31, 1915, 10.

2. For example, see Owen and Marie Elliot Interview Tape 1, "Dinsdale Community Oral History Project," Geoff Lewis, National Library of New Zealand, OHInt-0088/10, 1993. Or see ephemera such as "Electric Lamp House Limited: The Lamphouse, New Zealand's Leading Radio and

Electric Guide," photo by Electric Lamp House Ltd, New Zealand National Library, 1948, Eph-B-Radio-1948-01.

3. Marama Muru-Lanning, *Tupuna Awa: People and Politics of the Waikato* River (Auckland University Press, 2016), 94.

4. Linden Wilkinson, "Ten Foot Tall and Bullet-Proof,'—Assess, Adapt, and Overcome: The Anzac Legend in Modern Times," *Social Alternatives* 36, no. 2 (2017): 25–31; Michael Kilmister et al., "Treading on Sacred Ground? Confronting the Anzac Myth in Higher Education," *History Compass* 15, no. 8 (2017), https://doi.org/10.1111/hic3.12395.

5. Ministry of Works Statement (by the Hon. R. Semple, Minister of Works), *Appendix to the Journals of the House of Representatives*, 1948 Session I, D-01, 17.

6. E. O'Gorman et al., "Histories of Climate, Science, and Colonization in Australia and New Zealand, 1800–1945," *WIREs Climate Change* 7 (2016): 893–909.

7. James Lowe and Evan Parry, "Hydroelectric Schemes Compared with the Gas Industry," Appendix to the Journals of the House of Representatives, D-01b (Marcus F. Marks, 1918), 9–16.

8. Aaron Fox, "The Power Game: The Development of the Manapouri-Tiwai Point Electro-Industrial Complex 1904–1969" (PhD diss., University of Otago, Dunedin, 2001), 14.

9. Evan Parry, "Report by the Chief Electrical Engineer to the Hon. Sir W. Fraser, Minister of Public Works, Hydro-Electric Development: North Island Scheme," *AJHR*, 1918, D1A.

10. A. R. Shearer, *Report of the New Zealand Electricity Department*, Wellington, 1975; John F. Boshier, "New Zealand Electricity Supply," working paper, Massachusetts Institute of Technology, October 1976, 14.

11. Kingsley Baird, "The Past We Harvest That Was Yours: The Rhetoric of National Identity and the Legacy of the Unknown Warrior in New Zealand Memory," *Bulletin of the Auckland Museum* 21 (2020): 19–26; Katherine Smits, "Technologies of Nationalism: First World War Commemoration and New Zealand's *Gallipoli* Exhibition," *National Identities* 24 (2022): 271–86.

12. Miles Fairburn, "The Rural Myth and the New Urban Frontier: An Approach to New Zealand Social History, 1870–1940," *New Zealand Journal of History* 9, no. 1 (1975): 3–21; Bradford Haami, *Urban Māori: The Second Great Migration* (Oratia Media, 2018).

13. Jenny McArthur, "Auckland: Rescaled Governance and Post-

Suburban Politics," *Cities* 64 (2017): 79–87; Katie Higgins, "Lifestyle Migration and Settler Colonialism: The Imaginative Geographies of British Migrants to Aotearoa New Zealand," *Population, Space, and Place* 24 (2017), https://doi.org/10.1002/psp.2112.

14. Evelyn Stokes, "Tauponui a Tia: An Interpretation of Māori Landscape and Land Tenure," *Asia Pacific Viewpoint* 40 (2002): 137–58.

15. "Hydro-Electric Power," *New Zealand Herald*, June 21, 1927, 10, https://paperspast.natlib.govt.nz/newspapers/NZH19270621.2.91.

16. Malcolm McKinnon, *The Broken Decade: Prosperity, Depression and Recovery in New Zealand 1928–39* (Otago University Press, 2016), 35.

17. "Electric-Power Board and Supply Authorities Association Act, 1930," *GEO* 5, no. 42 (1930): 242–47.

18. "Power System: A Successful Year," *Evening Post*, September 12, 1939, 11.

19. There were over 37,000 lights, using over a million watts at the exhibition site. "The 1940 Centennial, Page 2—The Centennial Exhibition," New Zealand Ministry for Culture and Heritage, updated June 27, 2018, https://nzhistory.govt.nz/culture/centennial/centennial-exhibition.

20. "Current Events," Neuline Studios Ltd., Ngā Taonga Collections, F26292, 1945.

21. Helen Reilly, *Connecting the Country: New Zealand's National Grid, 1886–2007* (Steele Roberts, 2008), 108.

22. "Ministry of Works Act," no. 4 (1943).

23. "The Electricity Act," *Parliamentary Proceedings* (1945): 287–96.

24. Neil Rennie, *Power to the People: 100 Years of Public Electricity Supply in New Zealand* (Electricity Supply Association, 1989), 229.

25. "State Hydro-Electric Department Statement," *Appendix to the Journals of the House of Representatives*, 1948, Session I, D-04, p. 1.

26. "State Hydro-Electric Department Statement," p. 2.

27. Even though many questioned the safety and utility of widespread usage of SWER during the 1950s, it has become a mainstay for rural electric power distribution worldwide. Lloyd Mandeno (1888–1973), an electrical engineer, presented a paper to the New Zealand Institution of Electrical Engineers detailing the design's success in work for the Tauranga Electric Power Board in the 1920s. Lloyd Mandeno, "Rural Power Supply Especially in Back Country Areas," *Proceedings of the New Zealand Institution of Electrical Engineers* 33 (1947): 234.

28. Eric Pawson and Harvey Perkins, "Worlds of Wool: Recreating Value Off the Sheep's Back," *New Zealand Geographer* 69, no. 3 (2013): 208–20.

29. Reilly, *Connecting the Country*, 111.

30. Electricity and Gas Co-Ordination Act 1956, no. 106, pp. 1146–47.

31. Reilly, *Connecting the Country*, 115–19.

32. Ferdinand von Hochstetter, *New Zealand: Its Physical Geography, Geography and Natural History*, Edward Sauter, trans. (J. G. Cotta, 1867), 302–3.

33. "Electricity and Coal-Mines," *Appendix to the Journals of the House of Representatives*, session I-II, D-01A, pp. 5–6. Māori census data during this period is not reliable and tended to be taken by county or province rather than by town. "New Zealand 1916 Census," *Statistics New Zealand*, accessed July 26, 2023, https://www3.stats.govt.nz/historic_publications/1916-census/Report%20on%20Results%20of%20Census%201916/1916-report-results-census%20.html.

34. Reilly, *Connecting the Country*, 45.

35. D. Kear, "Potential Coal Sources for Steam-Electric Generation Plant Adjacent to Auckland," Coal Report 278, July 1951. Department of Scientific and Industrial Research, Wellington.

36. L. W. Gandar to Cabinet Works Committee, memorandum, "Office of the Minister of Electricity," June 26, 1972, Huntly Power Station Files, Archives New Zealand.

37. Atholl Anderson et al., *Tangata Whenua: A History* (Bridget Williams Books, 2014), 152.

38. "History of Māori Land," Ministry of Māori Development/Te Puni Kōkiri, last accessed August 31, 2023, https://www.tupu.nz/en/tuhono/about-maori-land-in-new-zealand/history-of-maori-land.

39. "Postmasters' Appointments," *Daily Southern Cross*, August 4, 1870, 3; "Untitled," *The Colonist* 19, July 26, 1877, 3, https://paperspast.natlib.govt.nz/newspapers/TC18770726.2.16; https://paperspast.natlib.govt.nz/newspapers/DSC18700804.2.22.

40. "Taupiri Mining Company," *New Zealand Herald*, August 10, 1877, 2, https://paperspast.natlib.govt.nz/newspapers/NZH18770810.2.18.

41. "New Zealand Coal as Fuel for Locomotives," *Appendix to the Journal of the House of Representatives*, session I, H-38, 1877, p. 7, https://paperspast.natlib.govt.nz/parliamentary/AJHR1877-I.2.2.3.42?items_per_page=10&query=%22Taupiri+Coal%22&snippet=true&sort_by=bySE.

42. Angela McCarthy, "'The Only Place Worth Thinking About': Personal Testimony and Irish Scottish Migrants in Australasia, 1921–1961," *Social History* 33 (2008): 317–35.

43. "Ralph's Taupiri Coal Mine: Official Visit of the Directors," *Auckland*

*Star,* August 12, 1898, 2, https://paperspast.natlib.govt.nz/newspapers/AS18980812.2.9.

44. "The Huntly Drivers," *Auckland Star,* February 4, 1914, 5, https://paperspast.natlib.govt.nz/newspapers/AS19130204.2.28.

45. John McCormick, *Reclaiming Paradise: The Global Environmental Movement* (Indiana University Press, 1911); Gregory Allen Barton, *Empire Forestry and the Origins of Environmentalism* (Cambridge University Press, 2004); Gary Brierley et al., "Re-Imagining Wild Rivers in Aotearoa New Zealand," *Land* 11, no. 8 (2022), https://www.mdpi.com/2073-445X/11/8/1272.

46. Faeeza Ballim, *Apartheid's Leviathan: Electricity and the Power of Technological Ambivalence* (Ohio University Press, 2023).

47. T. Buhrs, "Working Within the Limits: The Role of the Commission for the Environment in Environmental Policy Development in New Zealand" (PhD diss., University of Auckland, 1999), ii.

48. NZED, "Environmental Impact Report," October 1972, Huntly Power Station Files, Archives New Zealand.

49. J. A. Walding to the Prime Minister, memorandum, "Publication of Environmental Impact Reports," May 25, 1973, Huntly Power Station Files, Archives New Zealand.

50. NZED, *Environmental Impact Statement: Huntly,* October, 1972.

51. Minister of Finance to Michael Connelly, memorandum, "Huntly Power Station," May 28, 1973, Huntly Power Station Files 1972–1973, Archives New Zealand.

52. B. Powell to the Commissioner to the Environment, letter, "Environmental Impact Reports," October 1, 1974, Huntly Power Station 1973–1975, Archives New Zealand.

53. Robert Norman, *You Can't Win 'Em All: Confessions of a Public Works Engineer* (Slide Rule Press, 1997), 133.

54. "Public Right to Reports," *New Zealand Herald,* May 29, 1973, 2.

55. Robert Norman, *You Can't Win 'Em All: Confessions of a Public Works Engineer* (Slide Rule Press, 1997), 133–34.

56. Among the Waikato Tainui, he remains a complex figure, sometimes celebrated or disliked for his positions. Mahuta became well known in New Zealand history for his role in the Waikato Rauputu Claims Settlement Act (1995), sometimes called the Rauputu Claim, one of the first successful land claims that went through the Waitangi Tribunal, the court established to restore stolen Māori land. Furthermore, he actively engaged in the central land claims cases between the government and the Tainui during the 1940s.

Robert Mahuta, "Tainui, Kīngitanga and Rauputa," in *Justice & Identity*, ed. Margaret Wilson and Anna Yeatman (Bridget Williams Books, 1995), 25–28.

57. Robert Te Kotahi Mahuta, *Planning Issues*, 25.

58. Giselle Byrnes, "'A Dead Sheet Covered with Meaningless Words': Place Names and the Cultural Colonization of Tauranga," *New Zealand Journal of History* 36, no. 1 (2002): 18–35; Soma Chatterjee, "Immigration, Anti-Racism, and Indigenous Self-Determination: Towards a Comprehensive Analysis of the Contemporary Settler Colonial," *Journal for the Study of Race, Nation, and Culture* 25 (2019): 644–61.

59. As part of a larger volume on Māori history, Aroha Harris and Atholl Anderson wrote an extraordinary postscript about reasserting Māori place in history. Harris and Anderson, "The Past Matters," in *Tangata Whenua: A History* (Bridget Williams Books, 2015), 678. Increasingly in New Zealand's history education, scholars have pushed for greater Māori history and perspective in the curriculum—for example, Mark Sheehan, "Potentially Transformative: Aligning Māori Perspectives, Difficult Histories and Historical Thinking," in *Negotiating Ethnic Diversity and National Identity in History*, ed. Helen Mu Hung Ting and Luigi Cajani (Palgrave, 2023), 197–215.

60. Rodney Davies to P. J. Brooks, Commissioner for the Environment, "Environmental Assessment Studies," March 19, 1973, Huntly Power Station Files 1972–1973, Archives New Zealand.

61. Robert Te Kotahi Mahuta, *Planning Issues*, 4–5.

62. Robert Te Kotahi Mahuta, *Planning Issues*, 4.

63. Electricity Act 1968, New Zealand Legal Institute, accessed September 7, 2023, http://www.nzlii.org/nz/legis/hist_act/ea19681968n125166/.

64. Robert Mahuta, "Huntly Thermal Power Station—Comments on NZED Stage Two of Environmental Impact Report," June 1976, Huntly Power Station Submissions 1976, Archives New Zealand.

65. Marae Committee, "Submissions from Marae Committee, Waahi Pa, and the Communities at Waahi, TeKauri and Rakaumnga," Huntly Power Station Submissions 1976, Archives New Zealand.

66. Blakey (NZED) to Mahuta, "Huntly Power Station," March 3, 1973; Mahuta to Blakey (NZED), "Huntly Power Station," March 9, 1973, Huntly Power Station File 1972–1973, Archives New Zealand.

67. NZED, "Environmental Impact Report," October 1972, Huntly Power Station Files, Archives New Zealand.

68. Fisheries Conservation Act 1884, New Zealand Legal Institute, accessed September 6, 2023, http://www.nzlii.org/nz/legis/hist_act/

fca188448v1884n48330; Land Drainage Act 1893, New Zealand Legal Institute, accessed September 6, 2023, http://www.nzlii.org/nz/legis/hist_act/lda189357v1893n46176/.

69. For an in-depth analysis of such schemes, see Joseph Morgan Hodge, *Triumph of the Expert: Agrarian Doctrines of Development and the Legacies of British Colonialism* (Ohio University Press, 2007).

70. Evelyn Stokes, "Local Perception of the Impact of the Huntly Power Station 1971–1973," Occasional Paper 4, University of Waikato: Centre for Māori Studies and Research, 52.

71. Public Works Act 1928, New Zealand Legal Institute, accessed September 7, 2023, http://www.nzlii.org/nz/legis/hist_act/pwa192819gv1928n21231/.

72. Robert Norman to Chair of Officials Committee for the Environment to the Minister for the Environment, "Proposed Huntly Power Station Assessment of Environmental Impact," May 7, 1973, Huntly Power Station 1972–1973, Archives New Zealand.

73. Water and Soil Conservation Act 1967, New Zealand Legal Institute, accessed September 7, 2023, http://www.nzlii.org/nz/legis/hist_act/wasca19671967n135320/.

74. Town and Country Planning Act 1953, New Zealand Legal Institute, accessed September 7, 2023, http://www.nzlii.org/nz/legis/hist_act/tacpa19531953n91320/index.html.

75. National Water and Soil Conservation Authority, 1973, "Town and Country Planning Appeal Board in the Matter of Water Rights No. 220 and 221, July 6, 1973, Ministry of Justice Tribunals Unit," National Water and Soil Conservation Authority 1973, Archives New Zealand.

76. These appeals aligned with larger Māori movements in New Zealand, especially the hikoi, or Māori Lane March, in which protesters marched from Northland to Wellington between September and October 1975. The orchestrators and protesters of the hikoi, which reached some five thousand by the time they arrived in Wellington, sought to challenge the tenets of the Māori Affair Act (1953), which allowed the government to sell Māori land deemed "unproductive." Māori Affairs Act (1953), New Zealand Legal Institute, accessed September 7, 2023, http://www.nzlii.org/nz/legis/hist_act/maa19531953n94152/.

77. Aroha Harris, *Hīkoi: Forty Years of Māori Protest* (Huia, 2004), 10; Miranda Johnson, "'The Land of the Wrong White Crowd': Anti-Racist Organizations and Pakeha Identity Politics in the 1970s," *New Zealand Journal of History* 39, no. 2 (2005): 137–57.

78. For further examples of the ways in which Europeans colonized and invaded Waikato through both military intervention but also writing and photography, see Simon Dench, "Invading the Waikato: A Post-Colonial Review," *New Zealand Journal of History* 45, no. 1 (2011): 33–49.

79. P. Gavin, *The New Zealand Mining Handbook* (John Mackay, 1906), 396.

80. "Maori Lands: Vital Development Planned, Area of King Country," *Te Awamutu Courier* 79, no. 7126, October 31, 1949, 7.

81. Vincent Lagendijk, *Electrifying Europe: The Power of Europe in the Construction of Electricity Networks* (Aksant, 2008); Hudson Spivey, "The Technopolitics of Energy Transitions: Materiality, Expertise, and Fixed Capital in Japan's Power Grid Disputes," *Environment and Planning E: Nature and Space* 5, no. 3 (2022): 1106–22; Faeeza Ballim, *Apartheid's Leviathan: Electricity and the Power of Technological Ambivalence* (Ohio University Press, 2023).

82. "Two Inventions," *Auckland Star*, April 27, 1940, 4.

83. *Power from the River*, National Film Unit, Archive New Zealand, accessed June 25, 2025, https://www.youtube.com/watch?v=GmY52eeM2ZM.

84. Jarrod Hore, "Capturing Terra Incognita: Alfred Burton, 'Maoridom' and Wilderness in King Country," *Australian Historical Studies* 50, no. 2 (2019): 188–211.

85. *The Huntly Press*, March 14, 1973, 1.

86. Mahuta, *Planning Issues and the Huntly Power Station*, 4.

87. For example see, P. Nuttall and James E. Ritchie, New Zealand Tainui Māori Trust Board, University of Waikato Centre for Māori Studies and Research (University of Waikato, 1995). Since the founding of the Centre in 1972, faculty and administrators at the University of Waikato have expanded the program and renamed it to facilitate larger research initiatives. In 2010, it was restructured into the Te Kotahi Research Institute. More information can be found at "Who We Are," Te Kotahi Research Institute, University of Waikato, accessed June 1, 2026, https://www.waikato.ac.nz/research/institutes-centres-entities/institutes/tkri/about/.

88. Fookes, *Huntly Power Project: A Description* (Huntly Monitoring Project, University of Waikato, 1981), 16.

89. R. Fairclough, "An Overview of Power Generation in New Zealand," *Materials at High Temperatures* 24 (2007): 371–76.

90. "Coal to Make a Comeback as a Source of Energy," *Press*, January 21, 1985, 12; Alan Sherwood and Jock Phillips, "Coal and Coal Mining—The Late

20th and Early 21st Centuries," *Te Ara—The Encyclopedia of New Zealand*, http://www.TeAra.govt.nz/en/photograph/7439/huntly-power-station.

91. M. A. Chapman, "Human Impacts on the Waikato River System," *GeoJournal* 40 (1996): 85–99; M. Smith, "Culture Shock: The Legacy of the 1960s Power Generation Schemes in Aotearoa New Zealand," *Architectural History Aotearoa* 19 (2022): 61–82.

92. James Weir, "Second Huntly Unit Heads for Storage," *Stuff*, September 27, 2013, https://www.stuff.co.nz/business/industries/9218252/Second-Huntly-unit-heads-for-storage.

93. "Huntly Power Station," Genesis Energy, accessed September 7, 2023, https://www.genesisenergy.co.nz/about/generation/huntly-power-station.

94. Jamie Gray, "Genesis Completes Huntly Biomass Trial in Search for Alternatives to Coal, Gas," *New Zealand Herald*, February 21, 2023, accessed September 7, 2023, https://www.nzherald.co.nz/business/companies/energy/genesis-completes-huntly-biomass-trial-in-search-for-alternatives-to-coal-gas/5PVBCVC47RASTHBEP5KOHNP4CM/.

95. Olivia Wannan, "Huntly Burns Wood Instead of Coal During Short Trial," *Stuff*, February 22, 2023, https://www.stuff.co.nz/environment/climate-news/131298611/huntly-burns-wood-instead-of-coal-during-short-trial.

96. Greg Hurrell, "Stop Burning Coal at Huntly and Switch to Gas Now, Says New Energy Study," *Business Desk*, May 3, 2023, accessed September 8, 2023, https://businessdesk.co.nz/article/energy/stop-burning-coal-at-huntly-and-switch-to-gas-now-says-new-energy-study.

97. William Stanley Jevons, *The Coal Question: An Inquiry Concerning the Progress of the Nation, and the Probable Exhaustion of Our Coal-Mines* (Macmillan, 1865).

98. Jevons, *The Coal Question*, viii.

99. *Waikato Times*, March 29, 1973, 1.

100. *Evening Post*, September 12, 1977, 3.

## CHAPTER 5. "A PEBBLE IN YOUR SHOE"

1. Rawiri Waititi, "Unapologetic Māori Voice," *One News*, accessed October 27, 2023, https://www.1news.co.nz/2020/12/03/maori-party-promise-to-provide-unapologetic-maori-voice-in-parliament/.

2. Vincent O'Malley, *The Meeting Place: Māori and Pākehā Encounters, 1642–1840* (Auckland University Press, 2012), 227.

3. M. Durie, "Te Hoe Nuku Roa Framework: A Māori Identity Measure," *Journal of the Polynesian Society* 104 (1995): 461–70; Alistair Te Ariki

Campbell, *Māori Battalion: A Poetic Sequence* (Wai-te-ata Press, 2001); A. Gould, "From Taiaha to Ko: Repatriation and Land Settlement for Māori Soldiers in New Zealand After the First World War," *War and Society* 45, no. 2 (2009): 49–84; C. A. Houkamau et al., "The Multi-Dimensional Model of Māori Identity and Cultural Engagement," *New Zealand Journal of Psychology* 39, no. 1 (2010): 8–21.

4. Neil Rennie, *Power to the People: 100 Years of Public Electricity Supply in New Zealand* (Electricity Supply Association of New Zealand, 1987); John Martin, *People, Politics and Power Stations: Electric Power Generation in New Zealand, 1880–1990* (Bridget Williams Books and the Electricity Corporation of New Zealand, 1991); Helen Reilly, *Connecting the Country: New Zealand's National Grid, 1887–2007* (Steele Roberts, 2008).

5. Diana Montaño, *Electrifying Mexico: Technology and Transformation of the Modern City* (University of Texas Press, 2023)

6. Atholl Anderson et al., *Tangata Whenua: A History* (Bridget Williams Books, 2015), 226.

7. O'Malley, *Meeting Place*, 288.

8. Adele Manzella et al., eds., *Geothermal Energy and Society* (Springer, 2019), 194; "Energy in New Zealand 2023 Shows Renewable Electricity Generation Increased to 87%," Ministry of Business, Innovation, and Employment, accessed September 8, 2023, https://web.archive.org/web/20230912203647/https://www.mbie.govt.nz/about/news/energy-in-new-zealand-2023-shows-renewable-electricity-generation-increased-to-87-percent/.

9. Alexander Wyclif Reed, *Raupō Book of Māori Mythology* (Penguin, 2008).

10. Vincent O'Malley and David Anderson Armstrong, *The Beating Heart: A Political and Socio-Economic History of Te Arawa* (Huia, 2008).

11. Melissa Climo et al., "A History of Geothermal Direct Use Development in the Taupo Volcanic Zone, New Zealand," *Geothermics* 59 (2016): 215–24.

12. Waitangi Tribunal, *He Maunga Rongo—Report on the Central North Island Claims* (Wai 1200, Legislation Direct, 2008) vol. 4, chapter 20, 1494; Caren Fox, "Environmental Guardians: Learning from Māori Perspectives on Geothermal Fields," in *Oral History and the Environment*, Stephen Sloan and Mark Clave (Oxford University Press, 2022), 157–80.

13. Donald McLean, "Journal During a Visit to the Tribes of the Interior of the Northern Island of New Zealand," October 1845, MS-Group-1551, McLean Papers, Alexander Turnbull National Library.

14. "Death of Mr. Robert Graham," *New Zealand Herald*, May 27, 1885, 5.

15. "Waiwera Hot Springs," *Auckland Star*, June 5, 1875, 6.

16. Niramona Pini, "The Ohinemutu Hotel," *Bay of Plenty Times*, August 17, 1880, 3; "Police Court Ohinemutu," *Bay of Plenty Times*, October 7, 1880, 2; "A Trip on Wheels to the Lakes and Back," *Hawkes Bay Herald*, October 27, 1881, 3.

17. B. R. White and I. Chambefort, "Geothermal Development History of the Taupo Volcanic Zone," *Geothermics* 59 (2016): 149.

18. Ronan Foley et al., "Selling the Colonial Spa Town: The Contested Therapeutic Landscapes of Lisdoonvarna and Te Aroha," *Social and Cultural Geography* 44 (2011): 151–72.

19. Anderson et al., *Tangata Whenua*, 262.

20. Richard Boast, *Buying the Land, Selling the Land: Government and Māori Land in the North Island 1865–1921* (Victoria University Press, 2008), 41–42.

21. André Brett, *Acknowledge No Frontier: The Creation and Demise of New Zealand's Provinces, 1853–1876* (Otago University Press, 2016), 14–15.

22. Grant Morris, "James Prendergast and the Treaty of Waitangi: Judicial Attitudes to the Treaty During the Latter Half of the Nineteenth Century," *Victoria University of Wellington Law Review* 35, no. 1 (2004), http://www.austlii.edu.au/nz/journals/VUWLawRw/2004/4.html.

23. Alan Ward, "A 'Savage War of Peace'? Motives for Government Policies Towards the Kingitanga, 1857–1863," in *Raupatu: The Confiscation of Māori Land*, ed. Richard Boast and Richard S. Hill (Victoria University Press, 2010), 102.

24. Lindsay Cox, *Kotahitanga: The Search for Māori Political Unity* (Oxford University Press, 1993).

25. Boast, *Buying the Land*, 82–83; Steven Oliver, "Ngata, Paratene," *Dictionary of New Zealand Biography*, first published 1990, *Te Ara—the Encyclopedia of New Zealand*, accessed November 15, 2018, https://teara.govt.nz/en/biographies/1n7/ngata-paratene; M. P. K. Sorrenson, "Ngata, Apirana Turupa," *Dictionary of New Zealand Biography*, first published 1996, *Te Ara—the Encyclopedia of New Zealand*, accessed November 15, 2018, https://teara.govt.nz/en/biographies/3n5/ngata-apirana-turupa.

26. "Exploratory Report," *Ngati Porua Treaty Claims* (Waitangi Tribunal, 1993), 38.

27. *Weekly News*, May 26, 1952. Retrieved from Monty Soutar, "East Coast Places—Waiapu River Valley," *Te Ara—the Encyclopedia of New*

*Zealand*, accessed March 20, 2019, http://www.TeAra.govt.nz/en/east-coast-places/page-2.

28. "Native Trustee Act," 1920, New Zealand Legal Information Institute, accessed January 1, 2025, https://www.nzlii.org/nz/legis/hist_act/nta192011gv1920n21259/.

29. Richard Hill, *State Authority/Indigenous Autonomy: Crown-Maori Relations in New Zealand/Aotearoa* (Victorian University Press, 2005); Richard Hill, *Maori and the State: Crown-Maori Relations in New Zealand/Aotearoa* (Victoria University Press, 2010); Stella Coram, *Extinguishing Title: Maori Land Rights, People, and Perspective in the Post-Colonial New Zealand* (Common Ground Publishing, 2013); Marama Muru-Lanning, *Tupuna Awa: People and Politics of the Waikato River* (Auckland University Press, 2016).

30. "Census of the Population of the Dominion of New Zealand, October 15, 2016," Statistics New Zealand, Digitized Collection, accessed December 21, 2018, https://www3.stats.govt.nz/historic_publications/1916-census/Report%20on%20Results%20of%20Census%201916/1916-report-results-census%20.html?_ga=2.219019793.985914877.1546529967-789407428.1546529967.

31. Reilly, *Connecting the Country*, 123–26.

32. Lachy Paterson, "Hāwhekaihe: Māori Voices on the Position of 'Half-Castes' Within Māori Society," *Journal of New Zealand Studies* 9 (2010): 135–55; Paul Callister, "The Construction of Ethnicity and 'Belonging' in New Zealand: Where We Have Come from and Where We Might Be Going," *Journal of New Zealand Studies* 10 (2011): 115–38.

33. "Piako Council Meeting," *Te Aroha News*, January 21, 1920, 2; "Power and Light," *Putaruru Press*, March 14, 1940, 7.

34. George Graham, "Orakei Housing Conditions," *New Zealand Herald*, April 24, 1937, 17.

35. "'Twas Fifty Years Ago," *Evening Post*, January 21, 1890, 2.

36. "Duck Shooter Electrocuted," *Dominion News*, May 7, 1946, 7.

37. Christoper Gore's work, which is usefully open source, provides an excellent exploration of how the access to electric power can facilitate political power. Gore, *Electricity in Africa: The Politics of Transformation in Uganda* (James Currey, 2017).

38. David Christopher Young and Grid Heritage, "Interviews Abstracts," Māori Linemen Oral History Project, 2007–2008.

39. Rennie, *Power to the People*, 93.

40. D. Morrow and B. Brookes, "The Politics of Knowledge: Anthropol-

ogy and Māori Modernity in Mid-Twentieth Century New Zealand," *History and Anthropology* 24, no. 4 (2013): 453–71; R. Scott Sheffield and Noah Riseman, "Rehabilitating Assimilation," in *Indigenous Peoples and the Second World War: The Politics, Experiences and Legacies of War in the US, Canada, Australia, and New Zealand* (Cambridge University Press, 2018), 271–300.

41. Cathy Marr, *Public Works Takings of Māori Land, 1840–1981* (Waitangi Tribunal, 1997), 4–5; "What Is Māori Land," *Te Kooti Whenua Māori: Māori Land Court*, accessed June 3, 2026, https://www.māorilandcourt.govt.nz/en/maori-land/.

42. Boast, *Buying the Land*, 191–92.

43. Prince Ginori Conti, "The Larderello Natural Steam Power Plant," *Chemistry and Industry* 2 (1924): 739–41; Edmund Anscombe, "Nature's Boiler," *Evening Post*, October 2, 1945, 6.

44. "Geothermal Steam Act," *Appendix to the Journal of the House of Representatives*, August 25, 1952, 15.

45. Katherine Luketina and Phoebe Parson, in *Geothermal Energy and Society*, 193. The pause in construction between Wairakei and Kawerau was due to the success of the thermal power station, the Maui gas field expansion, and the environmental protection legislation introduced in the 1970s.

46. William Stanley Goosman, "Geothermal Power Survey," *Press*, February 22, 1950, 6.

47. See chapter 4 and "Lake Coleridge Power Station," Engineering New Zealand website, accessed June 3, 2026, https://www.engineeringnz.org/programmes/heritage/engineering-heritage-records/power-stations/lake-coleridge-power-station/.

48. K. Phair, "Direct Steam Geothermal Conversion Systems: Dry Steam and Superheated Steam Plants," *Geothermal Power Generation: Developments and Innovation*, R. DiPippo, ed. (Woodhead, 2016), 309.

49. C. J. Bromley et al., "Subsidence: An Update on New Zealand Geothermal Deformation Observations and Mechanisms," *World Geothermal Congress*, 1 (2015): 19–25.

50. Adele Manzella et al., *Geothermal Energy and Society* (Springer, 2019), 201.

51. Jo Whittle, "Into the Backyard: Huntly Power Station and the History of Environmentalism in New Zealand," *Australian and New Zealand Environmental History Network* 8, no. 1 (2013), accessed March 26, 2019, http://www.environmentalhistory-au-nz.org/2013/11/into-the-backyard-huntly-power-station-and-the-history-of-environmentalism-in-new-zealand/.

52. Matiu Rata Māori Affairs Amendment 1974.

53. W. H. Oliver, *Claims to the Waitangi Tribunal* (Department of Justice, 1991), 10; Andrew Sharp, *Justice and the Māori*, 2nd ed. (Oxford University Press, 1997), 74.

54. Nicola R. Wheen and Janine Hayward, "The Meaning of Treaty Settlements," in *Treaty of Waitangi Settlements*, ed. Nicola R. Wheen and Janine Hayward (Bridget Williams Books, 2012), 17.

55. Michael Belgrave, *Historical Frictions: Māori Claims and Reinvented Histories* (Auckland University Press, 2005) 1–2.

56. Robert Joseph Te Mata Hautū Taketake et al., *Te Pai Tawhiti: Exploring the Horizons of Māori Economic Performance Through Effective Collaboration* (Ngā Pae o Te Māramatanga, 2016), 13.

57. "The Taupo Railway," *Auckland Star*, October 2, 1914, 3.

58. Brian White and Isabelle Chambefort, "Geothermal Development History of the Taupo Volcanic Zone," *Geothermics* 59 (2016): 162.

59. "Te Ture Whenau Māori Act," 1993, accessed November 29, 2020, http://www.legislation.govt.nz/act/public/1993/0004/latest/DLM289882.html.

60. "Mokai Sleeps over N.Z. Hottest Water," *Press*, July 19, 1984, 15.

61. H. M. Bibby et al., "The Structure of the Moka Geothermal Field Based on Geophysical Observations," *Journal of Volcanology and Geothermal Research* 20, no. 1 (1984), 19; M. A. Grant and H. Barr, "Optimizing Field Proving and Development," *Conference Proceedings of the Tenth Workshop on Geothermal Reservoir Engineering*, January 21–22, 1985, 31.

62. "New Zealand Geothermal Fields," New Zealand Geothermal Association, accessed November 26, 2018, https://www.nzgeothermal.org.nz/geothermal-in-nz/nz-geothermal-fields/.

63. Sam Andrews, "Geothermal Taonga: In Trust at Mokai," *Mighty River Power Annual Report* (2002), 34.

64. Jessica McKenzie, "Showdown in Dixie Valley," *Bulletin of the Atomic Scientists*, November 2, 2022, accessed November 30, 2023, https://thebulletin.org/2022/11/geothermal-energy-dixie-valley-toad-biodiversity/: "History," Ormat Corporate Website, accessed November 30, 2023, https://www.ormat.com/en/company/welcome/history/#:~:text=Ormat%20was%20established%20in%201965,in%20the%20renewable%20energy%20sector.

65. Michael Dunstall, "Small Power Plants: Recent Developments in Geothermal Power Generation in New Zealand," *Geo-Heat Center Quarterly Bulletin* 20, no. 4 (1999): 5–12.

66. I do not have access to the original negotiations, conversations, or contracts between Ormat and the Tūaropaki Trustees. But the contracts

between Tūaropaki and Ormat illustrate what Tūaropaki valued and how the company designed the system around those values.

67. ORMAT, "Engineering, Procurement, and Construction Contract" (filed August 23, 2002), List of Exhibits, Exhibit A, "Scope of Work," 70; "Trustees of the Tuaropaki Trust," Māori Land Court Minute Book 77 (2002), https://www.courtsofnz.govt.nz/assets/benchbooks/mlcbb_public/#/home/14511/10/11

68. "Chronology of New Zealand Electricity Reform," Ministry of Business, August 2015, accessed December 2018, https://www.mbie.govt.nz/assets/2ba6419674/chronology-of-nz-electricity-reform.pdf.

69. Phil Rennie, "Gas a Big Opportunity for New Zealand," Petroleum Exploration and Production Association of New Zealand, November 15, 2017, https://www.energyresources.org.nz/news/global-demand-for-natural-gas-a-big-opportunity-for-new-zealand/.

70. "Mission Statement," Tūaropaki Trust, accessed January 8, 2019, http://www.tuaropaki.com/home/.

71. Emma Krasovich et al., "Harmonized Nitrogen and Phosphorus Concentrations in the Mississippi/Atchafalaya River Basin from 1980 to 2018," *Scientific Data* 9 (2022): 1–17.

72. "Tuaropaki Trust Wins Excellence in Farming Award," *SCOOP*, accessed January 19, 2019, http://www.scoop.co.nz/stories/BU0703/S00565/tuaropaki-trust-wins-excellence-in-farming-award.htm.

73. "About Us," Global Dairy Network, accessed January 8, 2019, https://www.globaldairynetwork.com/.

74. Waihaha Māori Lands Trust v Mokai Māori Trust—Tuaropaki A and Others [2018] Māori Land Court 191 Waiariki MB 234.

75. "Financial Report," Mercury New Zealand Limited (June 30, 2018), 30; Evelyn Stokes, *The Legacy of Ngatoroirangi: Māori Customary Use of Geothermal Resources* (University of Waikato, 2000), 93.

76. "Joint Development Agreement," Obayashi Website, accessed February 28, 2019, https://www.obayashi.co.jp/en/news/detail/news20190121_en_1.html.

77. Frederic E. Sautet, "Once Were IWI? A Brief Institutional Analysis of Maori Tribal Organisations Through Time," New Zealand Business Roundtable Working Paper, September 22, 2008; Maria Bargh, "The Triumph of Māori Entrepreneurs or Diverse Economies?" *Aboriginal Policy Studies* 1, no. 3 (2011): 53–69; Bargh, "Rethinking and Reshaping Indigenous Economies: Māori Geothermal Energy Enterprises," *Journal of Enterprising Communities* 6, no. 3 (2012): 271–83.

78. "Organizational Values," Tūaropaki Trust Website, accessed January 25, 2023, https://tuaropaki.com/our-story/organisation-values/.

79. "Transpower Integrated Report," August 28, 2023, accessed September 1, 2023, https://www.transpower.co.nz/news/transpower-announces-full-year-result-and-releases-integrated-report.

80. H. Legmann and D. Citrin, "First Twelve Months of Operation of the 60MW Mokai Geothermal Project, A High Pressure, Sustainable, and Environmentally Benign Power Plant," *Proceedings of the 5th INAGA Annual Scientific Conference*, March 7–10, 2001, 1–5.

81. Te Tai Ōhanga/New Zealand Treasury, "The Treasury Release of Submissions: Mixed Ownership Model Consultation with Māori Release Document," Release Document, March 2012, 1-5, accessed June 5, 2026, https://www.treasury.govt.nz/publications/information-release/submissions-received-extending-mixed-ownership-model-consultation-maori.

82. R. Bishop, "Initiating Empowering Research?," *New Zealand Journal of Educational Studies* 29, no. 1 (1994): 175–88; L. Pihama, "Tihei Mauri-ora, Honouring Our Voices, Manawahine as a Kaupapa Māori Theoretical Framework" (PhD diss., University of Auckland, 2001); Jo Mane, "Kaupapa Māori: A Community Approach," *MAI Review* 3 (2009): 1–9.

83. Linda Tuhiwai Smith, *Decolonizing Methodologies: Research and Indigenous People* (Palgrave, 2013), 292; Alison Jones, "Dangerous Liaisons: Pakeha, Kaupapa Māori, and Educational Research," *New Zealand Journal of Educational Studies* 47, no. 2 (2012): 100–112; Leonie Pihama et al., *Kaupapa Ranghau: A Reader* (Te Kotahi Research Institute, 2015).

84. Lachy Paterson and Angela Wanhalla, *He Reo Wāhine: Māori Women's Voices from the Nineteenth Century* (University of Auckland Press, 2017).

85. Merata Kawharu, "Kaitiakitanga: A Māori Anthropological Perspective of the Māori Socio-Environmental Ethic of Resource Management," *Journal of the Polynesian Society* 109, no. 4 (2000): 349–70; David Christopher Young and Grid Heritage, Māori Linemen Oral History Project, 2007–2008.

86. Merata Kawharu, "Kaitiakitanga: A Māori Anthropological Perspective of the Māori Socio-Environmental Ethic of Resource Management," *Journal of the Polynesian Society* 109, no. 4 (2000): 349.

87. Ian A. Thain, "Impact of the Resource Management Act on Future Geothermal Development in New Zealand," *Geothermics* 21, no. 5–6 (1992): 991–1000.

88. Mere Roberts et al., "Kaitiakitanga: Māori Perspectives on Conservation," *Pacific Conservation Biology* 2, no. 1 (1995): 7–20.

89. "Contact Energy, Ngati Tahu Win an International Award on Kenyan Geothermal Engagement Work," *ThinkGeo Energy*, April 20, 2018, accessed January 8, 2019, http://www.thinkgeoenergy.com/contact-energy-and-ngati-tahu-win-international-award-on-geothermal-engagement-work-in-kenya/.

90. "New Zealand Energy Scenarios: Navigating Energy Futures to 2050," *BusinessNZ Energy Council* (2015), 45.

91. "Te Ahi O Maui Geothermal Project," accessed January 8, 2019, http://www.eastland.nz/eastland-generation/projects/te-ahi-o-maui/.

92. Caitlin Smith et al., "The Sacred Nature of Geothermal Energy: How Power Africa Is Sharing Best Practices on Community Engagement Across Cultures and Continents," *Medium*, August 15, 2017, accessed January 8, 2019, https://medium.com/power-africa/the-sacred-nature-of-geothermal-energy-9f5f9f0de532.

## CONCLUSION: ELECTRIFICATION IS NOT A FINISHED PROCESS

*Epigraph*: William Pember Reeves, *New Zealand and Other Poems* (Grant Richards, 1898), 8

1. J. T. Salmon, *Heritage Destroyed: The Crisis in Scenery Preservation in New Zealand* (Reed, 1960), 67.

2. Tony Nightingale and Paul Dingwall, *Our Picturesque Heritage: 100 Years of Scenery Preservation in New Zealand* (New Zealand Department of Conservation, 2003), 56.

3. Carbon Disclosure Project, accessed March 23, 2023, https://web.archive.org/web/20230223101102/https://www.cdp.net/en/cities/world-renewable-energy-cities; "Energy Policy Review," International Energy Agency, April 2023, 9–14.

4. Aaron Fox, "The Power Game: The Development of the Manapouri-Tiwai Point Electro-Industrial Complex, 1904–1969" (PhD diss., University of Otago, 2001), 2–3.

5. "NZ Battery Project," Ministry of Business, Innovation, and Employment, accessed January 23, 2024, https://www.mbie.govt.nz/building-and-energy/energy-and-natural-resources/low-emissions-economy/nz-battery/.

6. W. E. Bardsley, "Pumped Storage for Reduced Emissions?" *New Zealand Hydrological Newsletter*, 4–6.

7. "Govt Confirms It Is Dumping 'Hugely Wasteful' Lake Onslow Battery Project," December 3, 2023, Radio New Zealand, https://www.rnz

.co.nz/news/national/503816/govt-confirms-it-is-dumping-hugely-wasteful-lake-onslow-battery-project; Richard Shaw, "Slow Down Simeon Brown—Bilingual Traffic Signs Aren't an Accident Waiting to Happen," *Stuff*, May 29, 2023, https://www.stuff.co.nz/opinion/300891998/slow-down-simeon-brown--bilingual-traffic-signs-arent-an-accident-waiting-to-happen.

8. Amanda Gillies, "Lake Onslow Pumped Hydro Back from the Dead," *Newsroom*, January 4, 2026, https://newsroom.co.nz/2026/04/01/lake-onslow-pumped-hydro-back-from-the-dead/.

# BIBLIOGRAPHY


## ARCHIVES CONSULTED

Archive New Zealand, Wellington (ANZ)
National Library of New Zealand, Wellington (NLNZ)
Lake District Museum
Te Papa
Waitangi Tribunal

### JOURNALS

*Appendix to the Journal of the New Zealand House of Representatives*
*California Journal of Mines and Geology*
*Electricity: A Popular Electrical Journal*
*Engineering and Mining Journal*
*Journal of the Institution of Electrical Engineers*
*Nature*
*New Zealand Journal of Science and Technology*
*Proceedings of the Electrical Engineering Advisory Committee*
*Proceedings of the New Zealand Institute*
*Telegraphic Journal and Electrical Review*
*The Electrician*
*The Electrical Engineer*
*The Electrical Journal*

*The Electrical World*
*Transactions of the Federated Institution of Mining Engineers*
*Transactions of the North of England Institute of Mining and Mechanical Engineers*
*Transactions of the Royal Society of New Zealand*

NEWSPAPERS

*Ashburton Guardian*
*Ashburton Times*
Auckland Star
Cromwell Argus
Evening Post
Grey River Argus
Inangahua Times
Kumara Times
Lake Wakatip Mail
Lyttleton Times
Nelson Examiner and New Zealand Chronicle
New Zealand Colonist
New Zealand Gazette
New Zealand Herald
New Zealand Times
Northern Advocate
North Otago Times
Otago Daily Times
Otago Witness
Taranaki Daily News
Thames Adviser
Thames Star
The Colonist
The New Zealand Journal
The Press
The Spectator
Timaru Herald
Tuapeka Times
West Coast Times
*Wellington Independent*

PUBLISHED SOURCES

NEWSPAPERS

*Ashburton Herald*
*Ashburton Times*

*Auckland Star*
*Bay of Plenty Times*
*Bush Advocate*
*The Colonist*
*Cromwell Argus*
*The Electrician*
*Evening Post*
*Evening Star*
*The Feilding Star*
*Gisborne Herald*
*Grey River Argus*
*Hawkes Bay Herald*
*Inangahua Times*
*Journal of Geomorphology*
*Kumara Times*
*Lake Wakatip Mail*
*Lyttleton Times*
*Manawatu Times*
*The Nelson Examiner*
*New Zealand Colonist*
*New Zealand Herald*
*New Zealand News*
*New Zealand Times*
*Otago Daily Times*
*Otago Witness*
*Parliamentary Debates*
*The Press*
*The Spectator*
*The Star*
*Sun*
*Thames Adviser*
*Thames Star*
*Timaru Herald*
*Tuapeka Times*
*Weekly Press*
*West Coast Times*

## SECONDARY SOURCES

Adas, Michael. *Dominance by Design: Technological Imperatives and America's Civilizing Mission*. Belknap Press, 2006.

Adas, Michael. *Machines as the Measure of Men: Science, Technology, and Ideologies of Western Dominance*. Cornell University Press, 2015.

## Bibliography


Aguiar, Marian. *Tracking Modernity: India's Railway and the Culture of Mobility*. University of Minnesota Press, 2011.

Agustoni, A., and M. Maretti. "Energy and Social Change: An Introduction." *International Review of Sociology—Revue Internationale de Sociologie* 22 (2012): 391–404.

Alessio, Dominic. "Promoting Paradise: Utopianism and National Identity in New Zealand, 1870–1930." *New Zealand Journal of History* 42, no. 1 (2008): 22–41.

Allen, R. C. "Backward into the Future: The Shift to Coal and Implications for the Next Energy Transition." *Energy Policy* 50 (2012): 17–23.

Andersen, Casper. *British Engineers and Africa, 1875–1914*. Pickering and Chatto, 2011.

Anderson, Atholl, Judith Binney, and Aroha Harris. *Tangata Whenua: A History*. Bridget Williams Books, 2015.

Anderson, David M., and David Killingray, eds. *Policing the Empire: Government, Control and Authority, 1830–1940*. Manchester University Press, 1991.

Anderson, Warwick. "Postcolonial Technoscience." *Social Studies of Science* 32, no. 5/6 (2002): 643–58.

Arnold, David. "Europe, Technology, and Colonialism in the 20th Century." *Technology and History* 21, no. 1 (2005): 85–106.

Arnold, David. *Everyday Technology: Machines and the Making of India's Modernity*. University of Chicago Press, 2013.

Arnold, Rollo. *New Zealand's Burning: The Settlers' World in the Mid-1880s*. Victoria University Press, 1994.

Aspden, R. "Electricity Supply in New Zealand: Origins." *IPENZ* 14, no. 2 (1988): 181–210.

Attard, B. "From Free-Trade Imperialism to Structural Power: New Zealand and the Capital Market, 1856–68." *Journal of Imperial and Commonwealth History* 35, no. 4 (2007): 505–27.

Attwood, B. "Law, History and Power: The British Treatment of Aboriginal Rights in Land in New South Wales." *Journal of Imperial and Commonwealth History* 42, no. 1 (2014): 171–192.

Badger, Wilfred. *Statutes of New Zealand: Whole Law of New Zealand, 1842–1893*. Croft-Rotherham, 1900.

Ballantyne, Tony. *Webs of Empire: Locating New Zealand's Colonial Past*. Bridget Williams Books, 2012.

Barak, On. *On Time: Technology and Temporality in Modern Egypt*. University of California Press, 2013.

Bargh, M. "Rethinking and Reshaping Indigenous Economies: Māori Geothermal Energy Enterprises." *Journal of Enterprising Communities* 6, no. 3 (2012): 271–283.

Bargh, M. "The Triumph of Māori Entrepreneurs or Diverse Economies?" *Aboriginal Policy Studies* 1, no. 3 (2011): 53–69.

Barker, R. "An Outline of the Mining History of New Zealand." *New Zealand Mining* 28 (2001): 36–43.

Barnes, Felicity. *New Zealand's London: A Colony and Its Metropolis*. University of Auckland Press, 2013

Barton, Ruth. "'Men of Science': Language, Identity, and Professionalization in the Mid-Victorian Scientific Community." *History of Science* 41 (2003): 73–119.

Belgrave, Michael. *Historical Frictions: Māori Claims and Reinvented Histories*. Auckland University Press, 2006.

Belich, James. *Making Peoples: A History of the New Zealanders from Polynesian Settlement to the End of the Nineteenth Century*. University of Hawaii Press, 2002.

Belich, James. *Paradise Reforged: A History of the New Zealanders from the 1880's to the Year 2000*. University of Hawaii Press, 2004.

Belich, James. *The Settler Revolution and the Rise of the Anglo-World, 1783–1939*. Oxford, 2009.

Belich, James. *The Victorian Interpretation of Racial Conflict: The Māori, The British, and the New Zealand Wars*. McGill-Queen's University Press, 1989.

Bell, Duncan S. A. "Dissolving Distance: Technology, Space, and Empire in British Political Thought, 1770–1900." *Journal of Modern History* 77, no. 3 (2005): 523–562.

Bennett, J. "Māori as Honorary Members of the White Tribe." *Journal of Imperial and Commonwealth History* 29, no. 3 (2001): 33–54.

Bennett, S. J. "Using Past Transitions to Inform Scenarios for the Future of Renewable Raw Materials in the UK." *Energy Policy* 50 (2012): 95–108.

Berthonnet, Arnaud. *Guide du chercheur en histoire de l'électricité*, Éditions La Mandragore. Paris, 2001.

Bijker, Wiebe E., Thomas Parke Hughes, and T. J. Pinch, eds. *The Social Construction of Technological Systems: New Directions in the Sociology and History of Technology*. MIT Press, 1987.

Bishop, R. "Initiating Empowering Research." *New Zealand Journal of Educational Studies, 29*, no. 1 (1994): 175–88.

Black, J. G. *Report on the Mining Industry of New Zealand*. George Didsbury, 1887.

Boast, Richard. *Buying and Selling the Land: Government and the Māori Land in the North Island, 1865–1921*. Victoria University Press, 2008.

Boast, Richard, and Richard S. Hill, eds. *Raupatu: The Confiscation of Māori Land*. Victoria University Press, 2010.

Boileau, Joanna. *Chinese Market Gardening in Australia and New Zealand: Gardens of Prosperity*. Springer, 2017.

Boubée, J., D. Jellyman, and C. Sinclair. "Eel Protection Measures Within the Manapouri Hydro-Electric Power Scheme, South Island, New Zealand." *Hydrobiologia* 609, no. 1 (2008): 71–82.

Brett, André. *Acknowledge No Frontier: The Creation and Demise of New Zealand's Provinces, 1853–1876*. Otago University Press, 2016.

Brien, P. K. "The Britishness of the First Industrial Revolution and the British Contribution to the Industrialization of Follower Countries on the Mainland, 1756–1914." *Diplomacy and Statecraft* 8 (1997) 48–67.

Britten, Rosemary. *Lake Coleridge: The Power, the People, The Land*. Hazard Press, 2000.

Broich, J. "Engineering the Empire: British Water Supply Systems and Colonial Societies." *Journal of British Studies* 46, no. 2 (2007): 346–365.

Bromley, C. J., S. Currie, S. Jolly, and W. Mannington. "Subsidence: An Update on New Zealand Geothermal Deformation Observations and Mechanisms." *World Geothermal Congress* (April 2015): 19–25.

Brooking, Tom., ed. *Making a New Land: Environmental Histories of New Zealand*. University of Otago Press, 2013.

Brooking, Tom. "Use It or Lose It: Unraveling the Land Debate in Late-Nineteenth Century New Zealand." *New Zealand Journal of History* 30, no. 2 (1996): 141–62.

Brooking, Tom, and Eric Pawson. *Seeds of Empire: The Environmental Transformation of New Zealand*. I. B. Tauris, 2010.

Buchanan, R. A. "Gentlemen Engineers: The Making of a Profession." *Victorian Studies*, 26 (1983): 407–29.

Bueltmann, Tanja. *Scottish Ethnicity and the Making of New Zealand Society, 1850–1930*. Edinburgh University Press, 2011.

Bunker, Raymond. "Systematic Colonization and Town Planning in Australia and New Zealand." *Planning Perspectives*, 3, no. 1 (2007): 59–80.

Burns, Patricia, and Henry Richardson, eds. *Fatal Success: A History of the New Zealand Company* (Auckland: Heinemann Reed, 1989).

Burton, Antoinette. *After the Imperial Turn: Thinking with and Through the Nation*. Duke University Press, 2003.

Burton, Antoinette. *The Empire in Question*. Duke University Press, 2011.

Byrnes, Giselle. *The New Oxford History of New Zealand*. Oxford University Press, 2009.

Cain, P. G. and A. G. Hopkins. *British Imperialism, 1688–2015*. 3rd ed. Routledge, 2016.

Calder, D. M. "Plant Ecology of Subalpine Shingle River-Beds in Canterbury, New Zealand." *Journal of Ecology* 49, no. 3 (1961): 581–94.

Campbell, Alistair Te Ariki. *Māori Battalion: A Poetic Sequence*. Wai-te-ata Press, 2001.

Cantor, Geoffrey, Gowan, Dawson, Graeme Gooday, et al. *Science in the*

*Nineteenth-Century Periodical: Reading the Magazine of Nature*. Cambridge University Press, 2004.

Carpenter, Lloyd. "Rich in Myth, Gold, and Narrative: Aspects of the Central Otago Gold Rush, 1862–2012." PhD thesis, Canterbury University, 2013.

Carse, A. "Nature as Infrastructure: Making and Managing the Panama Canal Watershed. *Social Studies of Science* 42, no. 4 (2012): 539–63.

Castells, Manuell. *The Rise of the Network Society*. Wiley-Blackwell, 1996.

Chakrabarty, Dipesh. *Provincializing Europe: Postcolonial Thought and Historical Difference*. Princeton University Press, 2007.

Chatterjee, Partha. *The Black Hole of Empire: History and the Global Practice of Power*. Princeton University Press, 2013.

Cherp, A., V. Vinichenko, J. Jewell, M. Suzuki, and M. Antal. "Comparing Electricity Transitions: A Historical Analysis of Nuclear, Wind and Solar Power in Germany and Japan." *Energy Policy* 10 (2017): 612–28.

Chibber, Vivek. *Postcolonial Theory and the Specter of Capital*. Verso, 2013.

Chikowero, Moses. "Subalternating Currents: Electrification and Power Politics in Bulawayo, Colonial Zimbabwe, 1834–1939." *Journal of African Studies* 33, no. 2 (2007): 287–306.

Christie, Renfrew *Electricity, Industry, and Class in South Africa*. State University of New York Press, 1984.

Clark, Geoffrey, Leach, Floss, and O'Connor, Susan. *Islands of Inquiry: Colonisation, Seafaring and the Archeology of Maritime Landscapes*. Acton, Australia: ANU Press, 2008.

Clayden, Arthur. *The England of the Pacific or New Zealand as an English Middle-Class Emigration-Field*. Wyman & Sons, 1879.

Clayden, Arthur. *The Revolt of the Field: A Sketch of the Rise and Progress of the Movement Among the Agricultural Labourers*. Hodder and Stoughton, 1874.

Climo, M., S. D. Milicich, and B. White. "A History of Geothermal Direct Use Development in the Taupo Volcanic Zone, New Zealand." *Geothermics* 59 (2016): 215–24.

Clive, Dewey. *Steamboats on the Indus: The Limits of Western Technological Superiority in South Asia*. Oxford, 2014.

Colley, Ann C. 2014. "Portrait, Empire, and Industry at the Belle Vue Zoo, Manchester." *Victorian Literature and Culture* 42 (2): 167–86.

Condevaux, A. "Māori Culture on Stage: Authenticity and Identity in Tourist Interactions." *Anthropological Forum* 19, no. 2 (2009): 143–61.

Cook, James. *A Voyage Towards the South Pole and Around the World*. W. Strahan and T. Cadell, 1777.

Cooper, Frederick. *Colonialism in Question: Theory, Knowledge, and Practice*. UCLA Press, 2005.

Cooper, Frederick, and Laura Stoler, eds. *Tensions of Empire. Colonial Cultures in a Bourgeois World.* University of California Press, 1997.

Cooper, Timothy. "Peter Lund Simmonds and the Political Ecology of Waste Utilization in Victorian Britain." *Technology and Culture* 52, no. 1 (2011): 21–44.

Coopersmith, Jonathan. *The Electrification of Russia, 1880–1926.* Cornell University Press, 1992.

Corn, Joseph J. *Imagining Tomorrow: History, Technology, and the American Future.* MIT Press, 1986.

Coutard, Olivier, ed., *The Governance of Large Technical Systems.* Routledge, 1999.

Crafts, N. "Productivity Growth in the Industrial Revolution: A New Growth Accounting Perspective." *Journal of Economic History* 64, no. 2 (2004): 521–35.

da Rosa, J. E. "Economics, Politics, and Hydroelectric Power: The Parana River Basin." *Latin American Research Review* 18, no. 3 (1983): 77–107.

Darmstadt, J., P. D. Teitelbaum, and J. G. Polach. *Energy in the World Economy: A Statistical Review of Trends in Output, Trade, and Consumption Since 1925.* Johns Hopkins University Press, 1971.

Darian-Smith, Kate, Patricia Grimshaw, and Stuart Macintyre. *Britishness Abroad: Transnational Movements and Imperial Cultures.* Melbourne University Press, 2007.

Darwin, John. *Unfinished Empire.* Bloomsbury, 2012.

Davey, C. J. "The Origins of Victorian Mining Technology, 1851–1900." *The Artefact* 19 (1996): 52–62.

David, T., & T. D. Dubois. "Local Religion and the Imperial Imaginary: The Development of Japanese Ethnography in Occupied Manchuria." *American Historical Review* 111, no. 1 (2006): 52–74.

De Silva, T. A., and S. L. Forbes. "Sustainability in the New Zealand Horticulture Industry." *Journal of Cleaner Production* 112 (2016): 2381–91.

Dieffenbach, Ernest. *New Zealand and Its Native* Population. Smith, Elder, 1841.

Dewey, Clive. *Steamboats on the Indus: The Limits of Western Technological Superiority in South Asia.* Oxford University Press, 2014.

DiPippo, Ronald, ed., *Geothermal Power Generation.* Woodhead Publishing, 2016.

Dirks, Nicholas B. *The Scandal of Empire: India and the Creation of Imperial Britain.* Belknap Press, 2006.

Dixon, Thomas. *Weeping Britannia: Portrait of a Nation in Tears.* Oxford University Press, 2015.

Downey, Gregory John, ed. *Taking Place: The Spatial Contexts of Science, Technology and Business.* Science History Publications, 2006.

Drayton, Richard. *Nature's Government: Science, Imperial Britain, and the 'Improvement' of the World*. Yale University Press, 2000.

Du Plessis, Rosemary, and Lynne Alice, eds. *Feminist Thought in Aotearoa New Zealand*. Oxford University Press, 1998.

Dunthorn, H., and M. J. Wintle, eds. *The Historical Imagination in Nineteenth-Century Britain and the Low Countries*. Brill, 2014.

Durie, M. "*Te Hoe Nuku Roa* Framework: A Māori Identity Measure." *Journal of the Polynesian Society* 104 (1995): 461–70.

Earp, George Butler. *The Emigrant's Guide to New Zealand: Comprising Every Requisite Information for Intending Emigrants, 5th edition*. W. S. Orr, 1849.

Eisenman, Harry James. *Charles F. Brush: Pioneer in Electrical Technology*. PhD thesis, Case Institute of Technology, 1967.

Eldred-Grigg, Steven. *Diggers, Hatter, and Whores: The Story of the New Zealand Gold Rushes*. Random House, 2008.

Evans, Julie. "'To Keep Within Proper Bounds . . .': Edward Eyre and the Colonised Peoples of Australia New Zealand, and the Caribbean." PhD thesis, University of Melbourne, 1998.

Evans, Julie, Patricia Grimshaw, David Philips, and Shurlee Swain. *Equal Subjects, Unequal Rights: Indigenous People in British Settler Colonies, 1830–1910*. Manchester University Press, 2003.

Fieldhouse, D. K. *Colonialism: 1870–1945: An Introduction*. Weidenfeld and Nicolson, 1981.

Fisher, Nick. "The Natural Theology of Victorian Industry." *Endeavour* 39, no. 1 (2015): 21–34.

Fei, Sheng, "Techniques Used by Chinese Immigrants in the Australian Gold Rushes and Their Influence." *Chinese Diaspora Studies* 6 (2013): 145–154.

Feinstein, C. H. "Pessimism Perpetuated: Real Wages and the Standard of Living in Britain During and After the Industrial Revolution." *Economic History* 58, no. 3 (1998): 625–58.

Fenton, F. D. "Observations on the State of the Aboriginal Māori Inhabitants of New Zealand." *Journals of the Statistical Society* (1860): 508–41.

Fjaestad, Mara. "Fast Breeder Reactors in Sweden: Vision and Reality." *Technology and Culture* 56, no. 1 (2015): 86–114.

Forster, M. E. "Hei Whenua Papatipu: Kaitiakitanga and the Politics of Enhancing Mauri Wetlands." PhD dissertation, Massey University, 2012.

Fouquet, Roger, and Peter J. G. Pearson. "Past and Prospective Energy Transitions: Insights from History." *Energy Policy* 50 (2012): 1–7.

Fox, Aaron. "The Power Game: The Development of the Manapouri-Tiwai Point Electro-Industrial Complex, 1904–1969." PhD thesis, University of Otago, 2010.

Frankema, Ewout. "Raising Revenue in the British Empire, 1870–1940" *Journal of Global History* 5, no. 3 (2010): 447–77.

French, J. M. *Native, Narrative, and Nation: The Construction of Self and Other in European Settler Colonies*. University of Chicago Press, 2001.

Friedel, Robert. *A Culture of Improvement: Technology and the Western Millennium*. MIT Press, 2007.

Fyfe, Aileen. *Steam-Powered Knowledge: William Chambers and the Business of Publishing, 1820–1860*. University of Chicago Press, 2012.

Gaffney, B. M. "What Price Water Marketing? California's New Frontier." *American Journal of Economics and Sociology* 56, no. 4 (1997): 475–520.

Geels, F. W. "The Multi-Level Perspective on Sustainability Transitions: Responses to Seven Criticisms." *Environmental Innovation and Societal Transitions* 1, no. 1 (2011): 24–40.

Geels, F. W. "Technological Transitions as Evolutionary Reconfiguration Processes: A Multi-Level Perspective and a Case Study." *Research Policy* 31, no. 8–9 (2002): 1257–74.

Geismar, H. "Resisting Settler-Colonial Property Relations? The WAI 262 Claim and Report in Aotearoa New Zealand." *Settler Colonial Studies* 3, no. 2 (2013): 230–243.

Gentry, Kynan. *History, Heritage, and Colonialism: Historical Consciousness, Britishness, and Cultural Identity in New Zealand, 1870–1940*. Manchester University Press, 2015.

Gibbons, Peter. "Cultural Colonization and National Identity." *New Zealand Journal of History* 36, no. 1 (2002): 5–17.

Gillbank, L. "The Origins of the Acclimatisation Society of Victoria: Practical Science in the Wake of the Gold Rush." *Historical Records of Australian Science* 6, no. 3 (1986): 359–74.

Gold, Barri J. *Thermopoetics: Energy in Victorian Literature and Science*. MIT Press, 2010.

Gooday, Graeme. *Domesticating Electricity: Expertise, Uncertainty and Gender, 1880–1914*. Pickering & Chatto, 2008.

Gooday, Graeme. *Morals of Measurement: Accuracy, Irony, and Trust in Late Victorian Electrical Practice*. University of Cambridge Press, 2004.

Gooday, Graeme, and Morris Low. "Technology Transfer and Cultural Exchange: Western Scientists and Engineers Encounter Late Tokugawa and Meiji Japan." *Osiris* 13 (1998): 99–128.

Gould, A. "From Taiaha to Ko: Repatriation and Land Settlement for Māori Soldiers in New Zealand after the First World War." *New Zealand Journal of History* 45, no. 2(2009): 49–84.

Grant, David. *Bulls, Bears, and Elephants: A History of the New Zealand Stock Exchange*. Victoria University Press, 1997.

Grant, Kevin, Philippa Levine, and Frank Trentmann, eds. *Beyond Sover-*

*eignty: Britain, Empire and Transnationalism, c. 1880–1950.* Palgrave Macmillan, 2007.

Greswell, William Parr. *History of the Dominion of Canada.* Clarendon Press, 1890.

Grey, George. *Polynesian Mythology and Ancient Traditional History.* John Murray, 1855.

Griffiths, J. "Were There Municipal Networks in the British World c. 1890–1939?" *Journal of Imperial and Commonwealth History* 37, no. 4 (2009): 575–97.

Grubler, A. "Energy Transitions Research: Insights and Cautionary Tales." *Energy Policy* 50 (2012): 8–16.

Gucciardo, Dorotea. "The Powered Generation: Canadians, Electricity, and Everyday Life." PhD dissertation, University of Western Ontario, 2011.

Gudgeon, Thomas Wayth. *The Defenders of New Zealand: Bring a Short Biography of Colonists Who Distinguished Themselves in Upholding Her Majesty's Supremacy.* H. Brett, 1887.

Halseth, Greg, ed., *Transformation of Resource Towns and Peripheries: Political Economy Perspectives.* Routledge, 2017.

Hamel, Jill. *The Archeology of Otago.* Department of Conservation, 2001.

Hannah, Leslie. *Electricity Before Nationalisation: A Study of the Development of the Electricity Supply in Britain to 1948.* Johns Hopkins University Press, 1979.

Harding, Sandra G. *The Postcolonial Science and Technology Studies Reader.* Duke University Press, 2011.

Harding, Sandra. *Science from Below: Feminisms, Postcolonialities, and Modernities.* Duke University Press, 2008.

Harris, A. "Dancing with the State: Māori Creative Energy and Policies of Integration, 1945–1967." PhD dissertation, University of Auckland. 2007.

Hasenöhrl, U. "Rural Electrification in the British Empire." *History of Retailing and Consumption* 4, no. 1 (2018): 10–27.

Hatch, Frederick H., and J. A. Chalmers. *The Gold Mines of the Rand.* Macmillan, 1895.

Hausman, William J., Peter Hertner, and Mira Wilkins. *Global Electrification: Multinational Enterprise and International Finance in the History of Light and Power.* Cambridge University Press, 2008.

Hayward, Janine, and Richard Shaw. *Historical Dictionary of New Zealand.* Rowman and Littlefield, 2016.

Headrick, Daniel R. *The Tentacles of Progress: Technology Transfer in the Age of Imperialism, 1850–1940.* Oxford University Press, 1988.

Headrick, Daniel R. *Tools of Empire: Technology and European Imperialism in the Nineteenth Century.* Oxford University Press, 1981.

Hecht, Gabrielle. *The Radiance of France: Nuclear Power and National Identity After World War II.* MIT Press, 1998.

Higgins, K. W. "Lifestyle Migration and Settler Colonialism: The Imaginative Geographies of British Migrants to Aotearoa New Zealand." *Population, Space and Place* 24, no. 3 (2018): 1–10.

Hill, David. *1788: The Brutal Truth of the First Fleet.* Random House, 2008.

Hill, Richard. *Māori and the State: Crown-Māori Relation in New Zealand/ Aotearoa, 1950–2000.* Victoria University Press, 2009.

Hill, Richard. "New Zealand Māori: The Quest for Indigenous Autonomy." *Ethnopolitics* 15, no. 1 (2016): 144–65.

Hilaire-Perez, L., and C. Verna. "Dissemination of Technical Knowledge in the Middle Ages and the Early Modern Era: New Approaches and Methodological Issues." *Technology and Culture* 47, no. 3 (2006): 536–65.

Hirsh, Richard F. *Power Loss: The Origins of Deregulation and Restructuring in the American Utility System.* MIT Press, 1999.

Hirsh, Richard F. *Technology and Transformation in the American Electric Utility Industry.* Cambridge University Press, 1989.

Hirt, Paul W. *Wired Northwest: The History of Electric Power, 1870s–1970s.* University of Kansas Press, 2012.

Hoag, Africa H. "Turning Water into Power: Debates over the Development of Tanzania's Rufiji River Basin." *Technology and Culture* 49, no. 3 (2008): 624–51.

Hochstetter, Ferdinand Ritter von. *New Zealand: Its Physical Geography, Geology, and Natural History.* J. G. Cotta, 1867.

Hodgins, Eric and Alexander Magoun. *Behemoth: The Story of Power.* Doubleday, Doran, 1932.

Hughes, Thomas P. *American Genesis: A Century of Invention and Technological Enthusiasm, 1870–1970.* Penguin, 1989.

Hughes, Thomas P. *Networks of Power: Electrification in Western Society, 1880–1930.* Johns Hopkins University Press, 1983.

Hughes, Thomas P. "The Seamless Web: Technology, Science, Etcetera, Etcetera." *Social Studies of Science* 61, no. 2 (1986): 281–92.

Hungerford, H., and S. Smiley. "Comparing Colonial Water Provision in British and French Africa." *Journal of Historical Geography* 52 (2016): 74–83.

Hunt, Bruce J. *The Maxwellians.* Cornell University Press, 1994.

Hunt, Bruce J. *Pursuing Power and Light: Technology and Physics from James Watt to Albert Einstein.* Johns Hopkins University Press, 2010.

Hursthouse, Charles. *New Zealand, "The Britain of the South."* Edward Stanford, 1861.

Ip, Manying, ed. *Unfolding History, Evolving Identity: The Chinese in New Zealand.* Auckland University Press, 2003.

Irwin, G. J. *The Prehistoric Exploration and Colonisation of the Pacific*. Cambridge University Press, 1992

Jäcksch, Hartmut, ed. *Māori und Gesellschaft: Wissensdchaftliche und literarische*. MANA-Verlag, 2007.

Jagger, Peter John, ed. *Gladstone*. Hambledon Press, 1998.

James, Harry. "Charles F. Brush: Pioneer Innovator in Electrical Technology." PhD dissertation, Case Institute of Technology, 1967.

Jasanoff, Sheila, and San-Huyn Kim. *Dreamscapes of Modernity: Sociotechnical Imaginaries and the Fabrication of Power*. Chicago University Press, 2015.

Johnson, J. C. F. *A Practical Treatise for Prospectors, Miners, and Students*. Charles Griffin, 1896.

Jones, Christopher F. 2013. "Building More Just Energy Infrastructure: Lessons from the Past." *Science as Culture* 22 (2): 157–63.

Jones, Christopher F. *Routes of Power: Energy and Modern America*. Harvard University Press, 2014.

Joseph, R., A. Tahana, J. Kilgour, et al. *Te Pai Tawhiti: Exploring the Horizons of Māori Economic Performance Through Effective Collaboration*. Māori Indigenous Governance Centre, University of Waikato, 2016.

Joseph, T. W. "Bulwer-Lytton's *The Coming Race* and an Anglo-Saxon Global 'Greater Britain.'" *Nineteenth-Century Contexts* 37, no. 3 (2015): 233–48.

Kaino, L. "There's Something Special About This Little Town': Cultural Identity and the Legacy of Hundertwasser in Kawakawa, New Zealand." *Continuum* 28, no. 1 (2014): 65–76.

Kale, S. S. 2014. "Structures of Power: Electrification in Colonial India." *Comparative Studies of South Asia, Africa and the Middle East* 34 (3): 454–75.

Kander, Astrid, Paolo Malanima, and Paul Warde, eds. *Power to the People: Energy in Europe over the Last Five Centuries*. Princeton University Press, 2014.

Kawharu, Ian Hugh. *Waitangi: Māori & Pākehā Perspectives of the Treaty of Waitangi*. Oxford University Press, 1989.

Kawharu, M. "Kaitiakitanga: A Māori Anthropological Perspective of the Māori Socio-Environmental Ethic of Resource Management." *Journal of Polynesian Society*, 109 (200): 349–370.

Kellow, A. *Transforming Power: The Politics of Electricity Planning*. Cambridge University Press, 1996.

Kelly, G. "History and Potential of Renewable Energy Development in New Zealand." *Renewable and Sustainable Energy Reviews* 15, no. 5 (2011): 2501–9.

Kelly, Morgan, Joel Mokyr, and Cormac Ó Gráda, "Precocious Albion: A New Interpretation of the British Industrial Revolution." *Annual Review of Economics* 6 (2014): 363–89.

Kim, Sang-Huyn and Sheila Jasanoff. "Containing the Atom: Sociotechnical

Imaginaries and Nuclear Power in the United States and South Korea." *Minerva* 47, no. 2 (2009): 119–20.

Kim, Sang-Huyn and Sheila Jasanoff. *Dreamscapes of Modernity: Sociotechnical Imaginaries and the Fabrication of Power.* University of Chicago Press, 2015.

King, Michael. *The Penguin History of New Zealand.* Penguin, 2003.

Knox, B. "Colonies and Colonisation in Bulwer Lytton's *The Caxtons, A Strange Story* and *The Coming Race.*" *Journal of Imperial and Commonwealth History* 44, no. 6 (2016): 857–80.

Landry, M. "Environmental Consequences of the Peace: The Great War, Dammed Lakes, and Hydraulic History in the Eastern Alps." *Environmental History* 20, no. 3 (2015): 422–48.

Larkin, B. "The Politics and Poetics of Infrastructure." *Annual Review of Anthropology* 42 (2013): 327–43.

Larnach, W. J. M. *The Handbook of New Zealand Mines.* George Didsbury, 1887.

Lagendijk, Vincent. *Electrifying Europe: The Power of Europe in the Construction of Electricity Networks.* Aksant, 2008.

Lambton, John George. *Report on the Affairs of British North America.* J. W. Southgate, 1839.

Lawrence, S., and P. Davies. "The Sludge Question: The Regulation of Mine Tailings in Nineteenth-Century Victoria." *Environment and History,* 20 (2014): 385–410.

Lee, Jung. 2014. "Invention Without Science: 'Korean Edisons' and the Changing Understanding of Technology in Colonial Korea." *Technology and Culture* 54 (4): 782–814.

Lester, A. "Imperial Circuits and Networks: Geographies of the British Empire." *History Compass* 4, no. 1 (2006): 124–41.

Lewis, M. "The Personal Equation: Political Economy and Social Technology on India's Canals, 1850–1930." *Modern Asian Studies* 41, no. 5 (2007): 967–94.

Lock, Alfred George Warnford. *Gold: Its Occurrence and Extraction.* E. & F. N. Spon., 1882.

Lorenzo, Veracini. *Settler Colonialism: A Theoretical Overview.* Springer, 2010.

Loveridge, S. "The 'Other' on the Other Side of the Ditch? The Conception of New Zealand's Disassociation from Australia." *Journal of Imperial and Commonwealth History* 44, no. 1 (2016): 70–94.

Luckin, Bill. *Questions of Power: Electricity and Environment in Inter-War Britain.* Manchester University Press, 1990.

Luis, N. "The Iron Industry Energy Transition." *Energy Policy* 50 (2012): 24–34.

Macfarlane, Daniel, and Peter Kitay. "Hydraulic Imperialism: Hydroelectric Development and Treaty 9 in the Abitibi Region." *American Review of Canadian Studies* 46, no. 3: 380–97.

Madan, Sandhya, Swetha Manimuthu, and S. Thiruvengadam. 2007. "History of Electric Power in India (1890—1990)." *2007 IEEE Conference on the History of Electric Power, HEP 2007*: 152–165.

Madureira, Nuno. L. "Oil in the Age of Steam." *Journal of Global History* 5, no. 1 (2010): 75–94.

Maganiello, C. J. "Hitching the New South to White Coal Water and Power 1890–1933." *The Journal of Southern History* 78, no. 2 (2012): 255–92.

Malm, Andreas. "Fleeing the Flowing Commons: Robert Thom, Water Reservoir Schemes, and the Shift to Steam Power in Early Nineteenth-Century Britain." *Environmental History* 19, no. 1 (2014): 55–77.

Malm, Andreas. *Fossil Capital: The Rise of Steam Power and the Roots of Global Warming*. Verso Books, 2016.

Mane, Jo. "Kaupapa Māori: A Community Approach." *MAI Review* 3 (2009): 1–9.

Manzella, Adele. Agnes Allansdottir, and Anna Pellizzone, eds. *Geothermal Energy and Society*. Springer, 2019.

Marcus, G. E. *Technoscientific Imaginaries: Conversations, Profiles, and Memoirs*. University of Chicago Press, 1995.

Marr, C. *Public Works Takings of Māori*. Waitangi Tribunal, 1997.

Marsden, Ben, and Crosbie Smith. *Engineering Empire: A Cultural History of Technology in Nineteenth-Century Britain*. Palgrave Macmillan, 2005.

Martin, John H. *People, Power and Power Stations: Electric Power Generation in New Zealand 1880–1990*. Bridget Williams Books and Electricity Corporation of New Zealand, 1991.

Mavhunga, Clapperton Chakanesta. *Transient Workspaces: Technologies of Everyday Innovation in Zimbabwe*. MIT Press, 2014.

Mayntz, Renate and Thomas P. Hughes, eds. *The Development of Large Technological Systems*. Campus Verlag, 1988.

McAloon, J. "By Which Standards? History and the Waitangi Tribunal: A Reply." *New Zealand Journal of History* 2, no. 2 (2006): 214–29.

McAloon, J. "Class in Colonial New Zealand: Towards a Historiographical Rehabilitation." *New Zealand Journal of History* 38, no. 1 (2004): 3–21.

McAloon, J. "Gentlemanly Capitalism and Settler Capitalists: Imperialism, Dependent Development and Colonial Wealth in the South Island of New Zealand." *Australian Economic History Review* 42, no. 2 (2002): 204–23.

McAloon, J. "Unsettling Recolonization: Labourism, Keynesianism and Australasia From the 1890s to the 1950s." *Thesis Eleven* 92, no. 1 (2008): 50–68.

McIntyre, David, ed. *The Journal of Henry Sewell*, vol. 2. Whitcoulls Publishers, 1980.

McIntyre, William David and W. J. Gardener, eds. *Speeches and Documents on New Zealand's History*. Clarendon Press, 1971.

McLean, Janet. "From Empire to Globalization: The New Zealand Experience." *Indiana Journal of Global Legal Studies* 11, no. 1 (2004): 161–81.

Meihana, P. N. "The Paradox of Māori Privilege : Historical Constructions of Māori Privilege Circa 1769 to 1940." PhD dissertation, Massey University, 2015.

Meiklejohn, J. M. D. *The British Empire: Its Geography, Resources, Commerce, Land-ways, and Water-Ways*. 6th ed. Alfred M. Holden, 1899.

Mein Smith, Philippa. *A Concise History of New Zealand*. Cambridge Concise Histories. Cambridge University Press, 2005.

Meiton, F. "The Radiance of the Jewish National Home: Technocapitalism, Electrification, and the Making of Modern Palestine." *Comparative Studies in Society and History* 57, no. 4 (2015): 975–1006.

Mikaere, Ani. *Colonising Myth, Māori Realities: He Rukuruku Whakaaro*. Huira Publishers, 2011.

Miller, Clark A., Alastair Iles, and Christopher F. Jones. "The Social Dimensions of Energy Transitions." *Science as Culture* 22, no. 2 (2013): 135–48.

Miller, Robert J., Jacinta Ruru, Larissa, Behrendt, and Tracy Lindberg. *Discovering Indigenous Lands: The Doctrine of Discovery in the English Colonies*. Oxford University Press, 2010.

Minami, Ryoshin. *Power Revolution in the Industrialization of Japan*. Kinokuniya, 1987.

Moon, Suzanne. *Technology and Ethical Idealism: A History of Development in the Netherlands East Indies*. CNWS Publications, 2007.

Mrazek, Rudolpf. *Engineers of Happy Land: Technology and Nationalism in a Colony*. Princeton University Press, 2002.

Morgan, Cecilia. *Building Better Britains: Settler Societies in the British World*. University of Toronto Press, 2016.

Morus, Iwan Rhys. 1998. *Frankenstein's Children: Electricity, Exhibition, and Experiment in Early 19th-Century London*. Princeton University Press, 1998.

Morus, Iwan Rhys. *Michael Faraday and the Electric Century*. Icon Books, 2004.

Morus, Iwan Rhys. "The Two Cultures of Electricity: Between Entertainment and Edification in Victorian Science." *Science and Education* 16 (2007): 593–602.

Mohsin, Anto. "Lighting 'Paradise': A Sociopolitical History of Electrification in Bali." *East Asian Science, Technology, and Society* 11, no. 1 (2016): 1–26

Mohsin, Anto. "Wiring the New Order: Indonesian Village Electrification and Patrimonial Technopolitics, 1966–1998." *Sojourn* 29, no. 1 (2014): 63–95.

Morrow, D. and M. Brookes. "The Politics of Knowledge: Anthropology and Māori Modernity in Mid-Twentieth Century New Zealand." *History and Anthropology* 24, no. 4 (2013): 453–71.

Murphy, Philip. *Monarchy and the End of Empire: The House of Windsor, the British Government, and the Postwar Commonwealth*. Oxford University Press, 2013.

Murton, G. "Making Mountain Places into State Spaces: Infrastructure, Consumption, and Territorial Practice in a Himalayan Borderland." *Annals of the American Association of Geographers* 107, no. 2 (2017): 536–45.

Myllyntaus, Timo. *Electrifying Finland: The Transfer of a New Technology into a Late Industrializing Economy*. ETLA, 1991.

Nagaoka, L. A. "Explaining Subsistence Change in Southern New Zealand: Foraging Theory Models." *World Archeology* 34 (2002): 84–102.

Nathan, S. "James Hector and the First Geological Maps of New Zealand." *Journal of The Royal Society of New Zealand* 44 (2014): 88–100.

Needham, Andrew. *Power Lines: Phoenix and the Making of the Modern Southwest*. Princeton University Press, 2014.

Neilson, G., G. Bignall, and D. Bradshaw. "Whakarewarewa, a Living Thermal Village–Rotorua, New Zealand." *Proceedings of the World Geothermal Congress*, Bali, Indonesia, April 2010, 25–29.

Nelson, C., and L. Grossenberg, eds. *Marxism and the Interpretation of Culture*. University of Illinois Press, 1988.

Nepe, Tukana Mate. "E hao nei e tneei reanga: Te Toi Huarwea Tupuna." Master's thesis, University of Auckland, 1991.

Nye, David. *The American Technological Sublime*. MIT Press, 1994.

Nye, David. *Consuming Power: A Social History of American Energies*. MIT Press, 1999.

Nye, David *Electrifying America: Social Meanings of a New Technology, 1880–1940*. MIT Press, 1992.

Nye, David. *When the Lights Went Out: A History of Blackouts in America*. MIT Press, 2010.

Oliver, W. H. *Claims to the Waitangi Tribunal*. Department of Justice, 1991.

O'Malley, Vincent. *The Meeting Place: Māori and Pākehā Encounters: 1642–1840*. Auckland University Press, 2012.

O'Sullivan, K. C., P. L. Howden-Chapman, and G. Fougere. "Making the Connection: The Relationship Between Fuel Poverty, Electricity Disconnection, and Prepayment Metering." *Energy Policy* 39, no. 2 (2011): 733–41.

Offer, Avner. "The British Empire, 1870–1914: A Waste of Money?" *Economic History Review* 46, no. 2 (1993): 215–38.

Orton, Harry. "The History of Underground Power Cables." *IEEE Electrical Insulation Magazine* 29, no. 4 (2013): 52–57.

Otter, Chris. 2007. "Making Liberal Objects." *Cultural Studies* 21 (4–5): 570–90.

Passer, Harold. *The Electrical Manufacturers, 1875–1900: A Study in Competition Technical Change, and Economic Growth*. Harvard University Press, 1953.

Penny, Guy. "Energy in Rural Māori Communities." *Water & Atmosphere* 13, no. 4 (2005): 18–19.

Pemberton, Robert. *The Happy Colony*. Saunders and Otley, 1854.

Petchey, Peter. *Bullendale: Phoenix Company's 1886 Electrical System Conservation Plan*. Southern Archeology, 2014.

Petchey, Peter. *Gold and Electricity Archaeological Survey of Bullendale*. Science and Technical Publishing, 2006.

Phillips, Gyllian. "Imaginary Africa and London's Urban Wasteland in Edith Sitwell's 'Gold Coast Customs.'" *Twentieth Century Literature* 56, no. 1 (2010): 71–91.

Phillips, J. "Verandahs and Fish and Chips and Footie on Saturday Afternoon: Reflections on 100 Years of New Zealand Historiography." *New Zealand Journal of History* 24, no. 4 (1990): 118–34.

Phimister, I. "Late Nineteenth-Century Globalization: London and Lomagundi Perspectives on Mining Speculation in Southern Africa, 1894–1904." *Journal of Global History* 10, no. 1 (2015): 27–52.

Pierce, Miles. "Early Electricity Supply in Melbourne." *Third Australasian Engineering Heritage Conference* IPENZ (2009): 1–10.

Pierre, Bill. *The North Island Main Trunk*. Reed, 1981.

Pietsch, Tamson. *Empire of Scholars: Universities, Network and the British Academic World 1850–1939*. Manchester University Press, 2013.

Pihama, Leonie. "Tihei Mauri Ora: Honouring Our Voices: Mana Wahine as a Kaupapa Māori Theoretical Framework." PhD thesis, University of Auckland, 2001.

Pollock, J. "Cultural Colonization and Textual Biculturalism: James Belich and Michael King's General Histories of New Zealand." *New Zealand Journal of History* 41, no. 2 (2007): 180–98.

Porter, A. "Introduction: Britain and the Empire in the Nineteenth Century." In *Oxford History of the British Empire, Volume 3, The Nineteenth Century*. Oxford University Press, 2014.

Pratt, Mary Louise. *Imperial Eyes: Travel Writing and* Transculturation. Routledge, 1992.

Price, A. Grenfell, ed. *The Explorations of James Cook in the Pacific: As Told by Selections of His Journals, 1768–1779*. Dover, 1971.

Priestley, Rebecca. *Mad on Radium: New Zealand in the Atomic Age*. Auckland University Press, 2012.

Pritchard, Sara B. *Confluence: The Nature of Technology and the Remaking of the Rhône*. Harvard University Press, 2011.

Pritchard, Sara B. "From Hydroimperialism to Hydrocapitalism: 'French' Hydraulics in France, North Africa, and Beyond." *Social Studies of Science* 42, no. 4 (2012): 591–615.

Putnam, P. C. *Energy in the Future*. Van Nostrand, 1953.

Railsback, L. Bruce. "Rain, Riches, and Empire: The Relationship Between Nations Ruling Distant Lands, Nations of Great Wealth, and Regions of Regular Moderate Atmospheric Precipitation." *Weather, Climate, and Society* 9, no. 3 (2017): 455–69.

Raugh, Harold E., ed. *The Victorians at War, 1815–1914: An Encyclopedia of British Military History*. ABC-Clio, 2004.

Reed, A. W. *Reed Book of Māori Mythology*. Reed, 2004.

Reed, Peter, Kate Schoonees, and Jeremy Salmond. *Historic Concrete Structures in New Zealand: Overview, Maintenance, and Management*. Science and Technology Publishing, 2008.

Reeves, Keir. "Sojourners or a New Diaspora: Economic Implications of Chinese Miners to the South-Pacific Goldfields." *Australian Economic History Review* 50, no. 2 (2010): 178–92.

Reeves, William Pember. *The Long White Cloud: Ao Tea Roa*. Horace Marshall & Son, 1898.

Reiger, Bernhard. *Technology and the Culture of Modernity in Britain and Germany, 1890–1945*. Cambridge University Press, 2005

Reilly, Helen. *Connecting the Country: New Zealand's National Grid 1886–2007*. Steele Roberts, 2008.

Rennie, Neil. *Power to the People: 100 Years of Public Electricity Supply in New Zealand*. Electricity Supply Association of New Zealand, 1989.

Rhodes, Richard. *Energy: A Human History*. Simon & Schuster, 2018.

Robert, L. "Pastoralism and the Transformation of the Rangelands of the South Island of New Zealand 1841 to 1912: Mt. Peel Station, a Case Study." PhD dissertation, University of Otago, 2007.

Roberts, Gwilym. "Sir John Anderson, 1814–86: The Unknown Engineer Who Made the British Empire Possible." *Transactions of the Newcomen Society* 78, no. 2 (2008): 261–91.

Roberts, M., W. Norman, N. Minhinnick, D. Wihongi, and C. Kirkwood. "Kaitiakitanga: Māori Perspectives on Conservation." *Pacific Conservation Biology* 2, no. 1 (1995): 7–20.

Rogers, D. "Oil and Anthropology." *Annual Review of Anthropology* 44 (2015): 365–80.

Rosa, E. A., and G. E. Machlis. "Energetic Theories of Society: An Evaluative Review." *Sociological Inquiry* 53, no. 2–3 (1983): 152–78.

Rosanowski, J. *Bottled Lightning: The Story of the Reefton Electric Light.* Cornwall & Associates, 2001.

Ross, C. "The Tin Frontier: Mining, Empire, and Environment in Southeast Asia, 1870s–1930s. *Environmental History* 19, no. 3 (2014): 454–79.

Rusden, G. W. *History of New Zealand.* Chapman and Hall, 1883.

Rutherford, J., and O. Coutard. "Urban Energy Transitions: Places, Processes and Politics of Socio-Technical Change." *Urban Studies* 51, no. 7 (2014): 1353–77.

Rydell, Robert, and Nancy Gwinn, eds. *Fair Representations: World Fairs and the Modern World.* Free University Press, 1994.

Said, Edward W. *Orientalism.* Vintage Books, 1978.

Sarfoh, Joseph A. *Hydropower Development in West Africa: A Study in Resource Development.* Lang, 1990.

Sargent, L. T. "Utopianism and the Creation of New Zealand National Identity." *Utopian Studies*, 12 (2001): 1–18.

Sargisson, Lucy, and Lyman Tower Sargent. *Living Utopias: New Zealand's Intentional Communities.* Ashgate, 2004.

Scheffield, R. Scott. *Indigenous Peoples and the Second World War: The Politics, Experiences and Legacies of War in the US, Canada, Australia, and New Zealand.* Cambridge University Press, 2018.

Schiffer, Michael B. "The Electric Lighthouse in the Nineteenth Century: Aid to Navigation and Political Technology." *Technology and Culture* 46, no. 2 (2005): 275–305.

Schiffer, Michael B. *Power Struggles: Scientific Authority and the Creation of Practical Electricity Before Edison.* MIT Press, 2008.

Schuurman, Frans J. "From Resource Frontier to Periphery: Agricultural Colonization East of the Andes." *Tijdschrift voor econmische en sociale geografie* 69, no. 1–2 (1978): 95–104.

Shamir, Ronen. *Current Flow: The Electrification of Palestine.* Stanford University Press, 2013.

Sharp, Andrew. *Justice and the Māori.* 2nd ed. Oxford University Press, 1997.

Sheng, Fei. "Techniques Used by Chinese Immigrants in the Australian Gold Rushes and Their Influence." *Chinese Southern Diaspora Studies* 6 (2013): 145–154.

Simon, Linda. *Dark Light: Electricity and Anxiety from the Telegraph to the X-Ray.* Harcourt Press, 2004.

Sinclair, Keith. *A Destiny Apart: New Zealand's Search for National Identity.* Nicholson Press, 1986.

Sinclair, Keith. *A History of New Zealand.* Penguin, 1959.

Singh, Mehta Uday. *Liberalism and Empire: A Study in Nineteenth-Century British Liberal Thought.* Chicago, 1999.

Smil, Vaclav. *Creating the Twentieth Century: Technical Innovations of 1867–1914.* Oxford, 2005.

Smil, Vaclav. *Energy in World History.* Westview Press, 1994.

Smil, Vaclav. *Transforming the Twentieth Century: Technical Innovations and their Consequences.* Oxford University Press, 2006.

Smith, Crosbie. *The Science of Energy: A Cultural History of Energy Physics in Victorian Britain.* University of Chicago Press, 1998.

Smith, Crosbie, and M. Norton. *Energy and Empire: A Biographical Study of Lord Kelvin.* Cambridge University Press, 1989.

Smith, Linda Tuhiwai, *Decolonizing Methodologies: Research and Indigenous People.* Palgrave, 2013.

Sovacool, B. K. "How Long Will It Take? Conceptualizing the Temporal Dynamics of Energy Transitions." *Energy Research and Social Science* 13 (2016): 202–15.

Speight, R. "Ice Wasting and Glacier Retreat." *Journal of Geomorphology* 3 (1940): 131–41.

Star, S. L. "The Ethnography of Infrastructure." *American Behavioral Scientist* 43, no. 3 (1999): 377–91.

Steer, Philip. "Gold and Greater Britain: Jevons, Trollope, and Settler Colonialism." *Victorian Studies* 58, no. 3 (2016): 436–63.

Stern, Philip J. *The Company-State: Corporate Sovereignty and the Early Modern Foundations of the British Empire in India.* Oxford, 2012.

Stokes, Evelyn, *The Legacy of Ngatoroirangi Māori Customary Use of Geothermal Resources.* University of Waikato, 2000.

Stokes, Evelyn. *Ohaaki: A Power Station on Māori Land.* Te Matahauariki Institute, University of Waikato, 2003.

Storey, Kenton. *Settler Anxiety at the Outposts of Empire: Colonial Relations, Humanitarian Discourse, and the Imperial Press.* UBC Press, 2016.

Storey, William Kelleher. *Guns, Race and Power in Colonial South Africa.* Cambridge University Press, 2008.

Storey, William Kelleher. *Science and Power in Colonial Mauritius.* University of Rochester Press, 1997.

Strauss, C. "The Imaginary." *Anthropological Theory* 6, no. 3 (2006): 322–44.

Tapine, Vyletta. *Building the Research Capacity Within Māori Communities: A Report on the Proceedings of the Wānanga.* New Zealand Council for Educational Research, 1999.

Taylor, C. "Modern Social Imaginaries." *Public Culture,* 14 (2002): 91–124.

Temin, P., and H. J. Voth. "Credit Rationing and Crowding Out During the Industrial Revolution: Evidence from Hoare's Bank, 1702–1862." *Explorations in Economic History* 42, no. 3 (2005): 325–48.

Thain, Ian A. "Impact of the Resource Management Act on Future Geothermal Development in New Zealand." *Geothermics* 21, no. 5–6 (1992): 991–1000.

Tokeley, A. H., and D. P. Brown. "Ohaaki Power Station: Geothermal Development in New Zealand." *Proceedings of the Institution of Mechanical Engineers* 206 (1992): 41–51.

Thompson, Silvanus. *Dynamo-Electric Machinery: A Manual for Student of Electrotechnics*. E.& F. N. Spon., 1886.

Tremewan, Peter. *The French at Akaroa: An Attempt to Colonize Southern New Zealand*. Canterbury University Press, 2010.

Trollope, Anthony. *Australia and New Zealand*. Bernhard Tauchnitz, 1873.

Tsing, A. "Resources Natural and Capitalist Frontiers." *Economic And Political Weekly* 38. no. 8 (2003): 5100–5106.

Tuffnell, S. "Engineering Inter-Imperialism: American Miners and the Transformation of Global Mining, 1871–1910." *Journal of Global History*, 10 (2015): 53–76.

Turnheim, B., and F. W. Geels. "Regime Destabilisation as the Flipside of Energy Transitions: Lessons from the History of the British Coal Industry, 1913–1997." *Energy Policy* 50 (2012): 35–49.

Tvedt, T. "Hydrology and Empire: The Nile, Water Imperialism and the Partition of Africa." *Journal of Imperial and Commonwealth History* 39, no. 2 (2011): 173–94.

Van der Vleuten, Erik. "Electrifying Denmark: A Symmetrical History of Central and Decentral Electricity Supply Until 1970." PhD diss., University of Aarhus, 1998.

Van der Vleuten, Erik. "Introduction: Networking Technology, Networking Society, Networking Nature." *History and Technology* 20, no. 5 (2004): 195–203.

Van der Vleuten, Erik, and Arne Kaijser, eds. *Networking Europe: Transnational Infrastructures and the Shaping of Europe, 1850–2000*. Science History Publications, 2006.

Viollet, P. L. "From the Water Wheel to Turbines and Hydroelectricity. Technological Evolution and Revolutions." *Comptes Rendus* 345, no. 8 (2017): 570–80.

Wagner, Tamara S., ed. *Domestic Fiction in Colonial Australia and New Zealand*. Pickering and Chatto, 2014.

Wahrman, D. "The Meaning of the Nineteenth Century: Reflections on James Belich's Replenishing the Earth" (Review). *Victorian Studies* 53, no. 1 (2010): 91–99.

Walker, Kathryn. "An English Empire of Law: Plantation Councils and the Atlantic Origins of the Imperial Constitution." PhD dissertation, New York University, 2016.

Walker, Ranginui. *Ka Whawhai Tonu Matou: Struggle Without End*. Penguin, 1990.

Ward, Alan. *An Unsettled History: Treaty Claims in New Zealand Today*. Bridget Williams Books, 2015.

Ward, J. A. "Early Railroad Empire Builders, 1850–1873." *Railroad History* 160 (1989): 5–21.

Ward, J. R. "The Industrial Revolution and British Imperialism, 1750–1850." *The Economic History Review* 47, no. 1 (1994): 44–65.

Warriner, Virginia Carolyn Ann. "Internationalisation of Māori Businesses in the Creative Industry Sector: Ko te rerenga o te toki a Tū, he whare orange." PhD thesis, Massey University, 2009.

Wenzlhuemer, Ronald. *Connecting the Nineteenth Century World: The Telegraph and Globalization*. University of Cambridge Press, 2013.

Wheeler, Nicola Rowan, and Janine Hayward, eds. *Treaty of Waitangi Settlements*. Bridget Williams Books, 2012.

Wheen, Nicola R., and Janine Hayward, eds. *Treaty of Waitangi Settlements*. Bridget Williams Books, 2012.

White, B. R., and I. Chambefort. "Geothermal Development History of the Taupo Volcanic Zone." *Geothermics* 59 (B) (2016): 148–67.

White, Richard. *Organic Machine: The Remaking of the Columbia River*. Hill and Wang, 1996.

Whittle, J. "Into the Backyard: Huntly Power Station and the History of Environmentalism in New Zealand." *ENNZ: Environment and Nature in New Zealand* 8, no. 1 (2013): 7–38.

Wilkenfeld, George. "The Electrification of the Sydney Energy System." PhD dissertation, Centre for Environmental and Urban Studies, Macquarie University, Sydney, 1989.

Williams, D. "Ko Aotearoa Tenei: Law and Policy Affecting Māori Culture and Identity." *International Journal of Cultural Property, 20* (2013): 311–31.

Wilson, C. "Up-Scaling, Formative Phases, and Learning in the Historical Diffusion of Energy Technologies." *Energy Policy* 50 (2012): 81–94.

Winks, Robin W. ed., *The Oxford History of the British Empire*, vol. 5. Oxford University Press, 2001.

Winseck, Dwayne R. and Robert M. Pike. *Communication and Empire: Media, Markets, and Globalization, 1860–1930*. Durham University Press, 2007.

Winskel, M. "When Systems Are Overthrown: The 'Dash for Gas' in the

British Electricity Supply Industry." *Social Studies of Science* 32, no. 4 (2002): 563–98.

Wood, G. A. "The Former Dominion of New Zealand." *Political Science* 26, no. 1 (1974): 2–10.

Woods, R. J. H. "From Colonial Animal to Imperial Edible: Building an Empire of Sheep in New Zealand, ca. 1880–1900." *Comparative Studies of South Asia, Africa and the Middle East*, 35, no. 1 (2015): 117–36.

Woollacott, Angela. *Settler Societies in the Australian Colonies: Self-Government and Imperial Culture*. Oxford University Press, 2015.

Wrigley, E. A., and R. S. Schofield. *The Population History of England 1541–1871*. 2nd ed. Arnold, 1981.

Wu, Shellen. *Empires of Coal: Fueling Chinas Entry Into the Modern World Order 1860–1920*. Stanford University Press, 2015.

Wu, Shellen. "The Search for Coal in the Age of Empires: Ferdinand von Richthofen's Odyssey in China, 1860–1920." *The American Historical Review* (2014): 339–63.

Yi-chong, Xu. *Sinews of Power: The Politics of the State Grid Corporation of China*. Oxford University Press, 2017.

Young, Frederick, ed. *Imperial Federation of Great Britain and Her* Colonies. S. W. Silver, 1876.

Young, Robert. *Empire, Colony, Postcolony*. Wiley, 2015.

Žmolek, Michael Andrew. *Rethinking the Industrial Revolution: Five Centuries of Transition from Agrarian to Industrial Capitalism in England*. Brill, 2013.

## ONLINE DATABASES

Archway, Archives New Zealand, https://collections.archives.govt.nz/#/

Dictionary of New Zealand Biography, https://www.teara.govt.nz/en/biographies

Engineering Heritage New Zealand, https://www.engineeringnz.org/programmes/heritage/

New Zealand Department of Conservation, https://www.doc.govt.nz/

Papers Past, New Zealand National Library, https://paperspast.natlib.govt.nz/

Statistics New Zealand, https://www.stats.govt.nz/

Te Ara—The Encyclopedia of New Zealand, https://teara.govt.nz/en

Trove Database, National Library of Australia, https://trove.nla.gov.au/

TrustPower, https://www.trustpower.co.nz/

# INDEX

*Note*: Page numbers in *italics* indicate figures, and references following "n" refer notes.


Adas, Michael, 38
agency, 73, 146. *See also* Māori
A. K. Smith's Carlton Foundry, 55
Alexander Turnbull Library, 21, 22
Allison, Charles, 93
aluminum resources, 168
*American Illuminations* (Nye), 9
Anderson, Atholl, 20, 205n59
Andrews, Sam Joesph, 155, 157
Anglo-American Brush Corporation, 69
Anglo-American Brush Electric Light Company, 69
Anglo Expansion, 28
*Apartheid's Leviathan* (Ballim), 117
Arapuni Dam, 132
Arapuni Power Station, 106, 147
archives, 21–23; as an artifact of colonialism, 21
Archives New Zealand, 21
Ardern, Jacinda, 3, 5, 172
Armstrong, William, 49
Arnold, David, 11
Arrow River, 59
Asseraf, Arthur, 11
*Auckland Star*, 26, 124
Australian and New Zealand Army Corps, 138
Aylmer's Race, 57
Ayrton, William, 69

Ballantyne, Tony, 19–20
Ballim, Faeeza, 117
Barak, On, 52
Bargh, Maria, 161
Barnes, Felicity, 19–20
Belich, James, 11, 19
Bickerton, Alexander, 26

Binney, Judith, 20
Binns, George J., 59, 61
biomass, 135
Birks, Lawrence, 93, 96, 97, 103
Bitchener, John, 113
Blacks Point, 32
Blakey, P. W., 124
Blyth, James, 49–50
Boast, Richard, 146
Britishness, 37, 38
Brooks, P. J., *121*, 122
Browne, Thomas Gore, 81
Brunton, Richard Henry, 69
Bullen, George F., 56–57, 60–61, 68–69
Buller, Charles, 12
Bureau of Land Reclamation, United States, 99–100

Canada, 78; hydroelectric infrastructures, 99; water rights, 99
Canterbury Frozen Meat and Dairy Produce Export Company, 92
Canterbury General Laborers' Union, 96
Cardwell, Edward, 82
Cardwell Reforms, 82
Carruthers, Joseph, 76
Centre for Māori Studies Research at the University of Waikato, 133
Chakrabarty, Dipesh, 52
Chatham Islands, 40
Childers, Hugh, 82
Childers Reforms, 82
Chinese immigrants, 71–72, 192n64
coal, 49, 61, 68, 74, 112; Jevons on, 135, 187n97; mines/mining, 104, 115–16
*The Coal Question* (Jevons), 135
Coates, Gordon, 107–8
Coleman, Leo, 38
Collins, William Whitehouse, 90
colonialism, 9, 16, 22, 38, 48, 72, 130, 139, 140; archives and, 21; concept of, 10; resources and, 8
Commonwealth of Australia, 84
Confederation and Annexation Bill, 83
Consolidated Zinc, 168
Constitution Act 1986, 110
Contact Energy Ltd., 150, 164
Cook, James, 40, 63, 72
Cook Strait Project, 111
Coram, Stella, 146
Counties Act, 131
Curley, Andrew, 8
Cyclone Gabrielle, 3

dairy, 107–8, 132, 145, 153, 160–61
Darwin, Charles, 47
Davies, Rodney, *121*, 122
Department of Māori Affairs, 145–46, 153, 155
Department of Scientific and Industrial Research (DSIR), 154
Depression years, 108–9
Dewey, Clive, 38
Dieffenbach, Ernst, 42
digitization, 22
District Commissioner of Works, 124
diverse economy, 161–62
Dobson, Arthur Dudley, 87, 89
"Doctrine of Discovery," 72
dominion, 76, 80–84: concept, 77; history, 77–78
du Fresne, Marc-Joseph Marion, 40

East Coast Māori Trust, 146
Eastern Group, 165
Edward, Frank, 102
electrical exhibitions, 26
Electrical Motive-Power Act, 86
electrical systems, 8–9
Electricity Act 1945, 109
Electricity Act 1968, 123, 126, 127
Electricity Act 2003, 157
Electricity and Gas Co-Ordination Act, 111
Electricity Commission, 157
Electricity Corporation of New Zealand (ECNZ), 154–55, 170
electricity/electric power: affordability, 109; centralization of production, 170; demand, 86, 107, 111–12; implementation, 26; schemes, 4, 109; Semple on, 109–10; a source of industrial power, 75. *See also specific power plant/station*
Electric Light and Power Company, 32
Electric Lines Act 1884, 85
Electric Power Boards and Supply Authorities Association Act 1930, 108
Electric Telegraph Act 1865, 85
*Electrifying America* (Nye), 9
Emergency Regulations Act 1939, 109
empire, concept of, 10
Emrod, 3; wireless electric power, 4
Energy Planning Committee of NZED, 113
energy policy, 164
energy security, 105, 109, 128, 134, 166
energy transition, 7–8, 20; consequences, 8; defined, 7
Environmental Council, 117
Environmental Defence Society (EDS), 118
Environmental Impact Report (NZED), 118–19, 123, 125
Environmental Protection and Enhancement Procedures (EPEP), 117–18
Escher Wyss, 87–88, 95, 97
Evans, Anthony Walton White, 69
*Evening Post*, 84, 91
export profits, 107
*Extinguishing Title* (Coram), 146

federalization, 83–84
Fisheries Conservation Act 1884, 125–26
FitzRoy, Robert, 47
Fletcher, Robert, 57–58, 69
Forest Act 1874, 44
fossil-fueled electric power production, 89, 170
Fox, William, 83
Fraser, William, 97, 112
Fuentes, Maria, 22

Gair, George, 135
garrison controversy, 82
Gavin, T., 113–14
Genesis Energy, 134
George, Ngaire, 160
Geothermal Combined Cycle technology (GCC), 156
geothermal power, 3–4, 18–19; dry steam facilities, 149; expansion/growth, 165–66; resources, 9, 141–42, 155, 157, 160. *See also* Tūaropaki Power Company
Geothermal Steam Act, 149

geothermal tourism industry, 142–43, 150
*Geothermics*, 162
Gladstone, William, 16, 39, 44–46, 50, 82–83
Global Dairy Network, 160
gold mines, 16–17, 30–31, 47, 57, 61, 67, 70, 71; as a colonial practice, 72; methods, 55. *See also* Phoenix Gold Mine
gold rush, 28, 47, 62, 65–66; California, 28, 55; Otago, 55, 66, 71–72
Goosman, W. H., 149
Gourmet Mokai, 159–60
Gourmet Paprika Ltd., 159
Graham, David, 38
Graham, George, 147
Graham, Robert, 142
Grand Centennial Exhibition, 109
green hydrogen, 3
Grey, George, 25, 42–43, 45, 81, 83
*Grey River Argus*, 37

Hancock, L. M., 91
*The Handbook of New Zealand Mines* (Larnach), 70
Hård, Mikael, 102
Harris, Aroha, 20, 205n59
Hauhungaroa Partnership, 159
Hawaiki, 141
Hay, P. S., 91
Hay, Robert, 89
*Heritage Destroyed* (Salmon), 167, 168
Hiku, Bill, 148
Hill, Richard, 20, 146
Hobson, William, 43
Hochstetter, Ferdinand Ritter von, 66, 112, 135
Holmes, R. W., 93
Horahora Power Station, 113
Hore, Jarrod, 132
Huntly: City Council, 113–14; extractive industry and, 114; water-rights legislation, 104–5. *See also* Huntly Power Station
Huntly Power Station, 17–18, 103–36, 154, 168, 170; as a backup power source, 135; boilers, 134–35; as largest thermal power station, 104; location, 122, *123*; Mahuta and, *104*, 104–5, 120, 122, 124, 128–30, 132–33, 135; planning, 105; proposal, 113; steam turbines, 104
*The Huntly Press*, 133
Hursthouse, Charles, 16, 44, 46–47, 64, 81; *New Zealand or Zealandia*, 46, 64, 81
Hydroelectric Branch of PWD, 103, 109
hydroelectricity/hydropower, 4, 17, 48–49, 67, 73, 150–51; Coates on, 107–8; colonialization and, 72; engineering community, 74; expandability, 111; Parry on, 106; river system, 74. *See also* Lake Coleridge Power Station; Phoenix Gold Mine
hydroimperialism, 98–102, 170

Immigration and Public Works Act, 85
Imperial Federation League, 84
imperialism, 38, 44, 83; concept, 10
*Inangahua Herald*, 29
Inangahua River Valley, 27, 29
*Inangahua Times*, 29

India, 10, 42, 99
Indonesian Geothermal Association, 162
Ip, Manying, 20

Jameson, Leander Starr, 76
Jasanoff, Sheila, 38–39
JASMaD, *121*, 122
Jevons, William Stanley, 49, 135, 187n97

Kaiapoi Woollen Mills, 87
kaitiakitanga, 164
Kaupapa Māori, 21, 163
Kawerau Ahu Whenua Trust, 165
Kennan, Danny, 19–20
Kenya Electricity Generating Company (KenGen), 165–66
Kidd, Benjamin, 71
Kidston, William, 76
Kīngitanga (Māori King) Movement, 112, 114–15, 122, 131, 143

labour kingdom, 12
Labour Party, 3, 111, 125, 137, 151
Lake Coleridge Power Station, 17, 80, 87, 89–102, *79*, *100*, 170
Lake Onslow, *171*, 172–73
Land Development Committee, 52
Land Drainage Act 1893, 125, 126
Land Grants, 114
Land Trusts (Māori), 140–41, 144–48, 166
Lemon, Charles, 26
*Lyttleton Times*, 49, 88, 98

Maasai, 165–66
Mahuta, Robert Te Kotahi, 104–5, *104*, 120, 122, 124, 128–30, 132–33, 135
Mandeno, Lloyd, 202n27
Mangahao-Waikaremoana, 109
Māori, 3–5, 23, 39–44, 73, 107, 146, 148, 163; agency, 73, 146; archival absence, 22; business models, 144; California gold rush, 55; colonial power dynamics, 12; employment and income, 145; equal representation, 144; gold rush and, 71; hydroelectric systems and, 150–51; integration, 152; kaitiakitanga, 164; land confiscation, 40, 43, 114, 122, 125, 129, 131, 133; land holdings, 145; land owners, 43–44, 72, 126, 133, 143, 152–53, 166; land tenure, 144; Land Trusts, 140–41, 144–48, 166; New Zealand Wars (*see* New Zealand Wars); politicians and community leaders, 144; Rēkohu, 40; resisting settler colonialism, 138; revitalization, 145; rights and interests, 145; rights movements, 170; studies programs, 21; trade and traders, 73; water rights, 100; world wars and, 138. *See also* Tūaropaki Power Company; Tūaropaki Trust
Māori Affairs Amendment Act 1974, 151
*Māori and the State* (Hill), 146
Māori Cultural Renaissance, 129, 151
Māori Jack, 71
Māori King Movement. *See* Kīngitanga (Māori King) Movement

Māori Land Act. *See* Te Ture Whenua Māori Act
Māori Land Development Scheme, 132, 144
Māori Musket Wars, 73
Māori Parliament, 144
Māori Representation Act, 144
Massey, William F., 101
McKenzie, Roderick, 93, 94
McLeod, Norman, 126
Meiklejohn, John Miller Dow, 99
Meiton, Fredrik, 5
Meremere (electrification project), 154
"Messrs. Kohn & Co," 25
Mighty River Power, 157
Ministry of Works (MoW), 105, 110–11, 113, 117–20, 123–30, 132–35, 148–49, 154
Minogue, Michael, 135
Miraka Dairy, 160
Mōkai, 152–54; geothermal fields, 154–55
Montaño, Diana, 139
Mosgiel Woolen Mill, 58
Municipal Corporations Act 1886, 85
Murray Creek, 27
Muru-Lanning, Marama, 146

Napier Earthquake, 108
National Film Unit, 132
National Library, 104
National Parks Authority, 167
National Party, 111–12, 172
National Water and Soil Conservation Authority, 129
Native Department, 125, 145–46. *See also* Department of Māori Affairs
Native Lands Act, 143, 145
Native Lands Courts, 23, 143
Native Trustee Act, 145
natural disasters, 108
Nature Conservation Council, 117
Nature Conservation Council Act 1962, 117
Nectar, 150
Needham, Andrew, 5
Nepe, Tukana Mate, 21
newspapers, 27
New Zealand, 10, 22, 37, 107; as Britain of the South, 37, 39, 45; Depression/Depression years, 108–9; as epitome of a colonial frontier, 169; export profits, 107; historiography, 19–23; as a "mineral kingdom," 47; as a model settler-colony, 62–63, 71; national electric utility, 17; population, 73, 107; Second Constitution, 44; as a site of promise, 12, 14; sub-imperialism, 84; as a sustainable and green country, 168; transition to dominion, 80–84; Trilemma rating index, 164–65; visions, 44–50
New Zealand and Polynesian Company, 83
New Zealand Battery Project, *171*, 172, 173
New Zealand Constitution Amendment (Request and Consent) Act 1947, 110
New Zealand Electrical Construction Company, 89
New Zealand Electricity Department (NZED), 105, 113, 116–17, 128, 130, 149, 151; Energy Plan-

ning Committee, 113; Environmental Impact Report, 118–19, 123, 125; formation, 104; Waahi O Marae Committee, 124; Waahi Pā Marae, 122–24; water rights, 127, 129
New Zealand Electric Light Company, 69
*New Zealand Graphic*, 31
*New Zealand Herald*, 119
New Zealand Institute, 47
*The New Zealand Journal of History*, 21
*New Zealand or Zealandia* (Hursthouse), 46, 64, 81
New Zealand Settlements Act 1863, 114
*New Zealand Times*, 49
New Zealand Wars, 11, 43, 72, 81–82, 84–85, 112, 114–15, 143, 147
Ngaire George Sustainability Centre, 160
Ngarewa-Packer, Debbie, 137
Ngata, Āpirana, 144–45
Ngata, Paratene, 144
Ngāti-Porou Dairy Company, 145
Ngāti Tahu, 150, 164
Ngāti Toa, 143
Ngātoroirangi, 141
Ngawha Power Station (10 MW), 156
Niagara Falls, 49, 89
Norman, Robert, 118–19, 127
Northern King Country, 132
North Island, 9, 17, 96, *104*, 106, 146–47; Coromandel Peninsula, 66; land losses on, 143; mines, 67; Napier Earthquake, 108; New Zealand Wars, 84; power for, 111–16. *See also* Huntly Power Station
North Island Main Trunk, 92
*Nova Zeelandia*, 40
Nye, David, 9

Obayashi Corporation, 161
Ōhakuri Dam, 150
oil crisis, 127, 170
Oliver, Bill, 151
Olkaria Power Station, Kenya, 165
O'Malley, Vincent, 19
Ōrākei Kōrako geothermal system, 150
Orakei village, 147
Ormat Pacific, 155–56
Ormat Technologies, 165
Otago Gold Rush, 55, 66, 71–72
Otago School of Mines in Dunedin, 69
overpopulation: Malthusian fears, 12

Pacific Federation, 83
Parihaka (Māori settlement), 147
Parry, Evan, 96, 106, 112–13
Paterson, Lachy, 163
pauperism, 12
Pemberton, Robert, 12, *13*
Petchey, Peter, 58, 61, 62
Phoenix Gold Mine, 17, 54–75, 169; Binns on, 59, 61; Bullen, 56–57, 60–61, 68–69; economic development, 70; powerhouse locations, 57, *58*; transporting coal, 61
Phoenix Quartz Mining Company, 56

Physical Environment Committee, 117
Poihipi Power Station (55 MW), 156
Polynesian peoples, 40
Powell, B. G., 118
Pratt, Mary, 11
Prendergast, James, 143
Prince, Walter, 31–36, 48–50, 52–53; Anglo-American Brush Corporation, 69; Phoenix Quartz Mining Company, 56; R. E. Fletcher and Co., 32, 57
Public Works Act 1870, 85
Public Works Act 1928, 126
Public Works Act 1981, 127
Public Works Act Amendment Act 1889, 131
public works agencies, 170
Public Works Amendment Act 1908, 101, 113
Public Works Annual Report, 97
Public Works Department (PWD), 17–18, 170; centralization of power production, 105; effectiveness, 101; expansion, 102; foundation, 85; geothermal energy for, 148; Grand Centennial Exhibition, 109; Hydroelectric Branch, 103, 109; infrastructures, 102; Manapōuri Hydroelectric Plant, 168
Putaruru-Mōkai railway, 153

quartz fields, 29
Quartzopolis. *See* Reefton

Railways Construction and Land Act 1881, 131
Rakaia River, 91, 95
Rakaumanga School, 124–25
Ralph, Anthony, 115
Rata, Matiu, 151
Reefton, 15, 16, 18, 25–53, *31*, *33*; cosmopolitan environment, 28; engineering and technological debates, 29–30; gold/gold industry, 30, 31; international fame, 52; map of hydroelectric scheme, *28*; population growth, 28, 53; Prince, 31–36, 48–50, 52–53, 57; quartz, 27–28, 31; settlers building, 27; stock exchange, 29
Reefton Electric Light Company, 27
Reeves, William Pember, 167
Reilly, Helen, 101
renewable electricity projects, 3
renewable energy/power, 168
Rennie, Neal, 101
resources, colonialism and, 8
Rhodes, Heaton, 98
Rotokawa Power Station (24 MW), 156
Ruatōria Factory, 145
Rudy, Jason, 38
Rural Electricity Reticulation Council, 110
Ruskin, John, 49
Ryan, James, 95

Salmon, J. T., 167, 168
Sautet, Frédéric, 161
Scandinavian Quartz Mining Company, 55
Scandinavian Reef, 55
Seddon, Richard, 86, 98, 116
self-governing colonies, 77–78, 83
self-sufficiency, 5, 14, 79, 80, 98, 102, 105, 136, 158, 169

Semple, Robert, 109–10
settler capitalists, 29
settler-colonialism, 4–6, 7, 17, 29, 39, 42, 46, 77, 78, 102, 138, 172; concept, 10; Dominion, 170; hegemonic practices, 59, 62, 71, 158; modern systems and, 169; subject-colonialism *vs.*, 10. *See also* Māori
Sewell, Henry, 82–83
Sharp, Andrew, 151
Siemens, C. W., 48
Siemens, William, 49
Simmons, Peter Lund, 71
Single Wire Earth Return (SWER), 110–11, 202n27
Skipper's Creek, 55, *60*
Smith, Graham, 163
Smith, Linda Tuhiwai, 19, 22, 163
sociotechnical imaginary, 38–39
Soil Conservation and Rivers Control Act 1941, 126
Southern Island Transmission Network, 168
South Island, 40, 51, 55, 84–85; Canterbury region, 92; gold mines, 66, 67. *See also* Lake Coleridge Power Station
Spencer, Herbert, 9
Sprague Company, 26
SS *Manapouri*, 25–26
Stafford, Edward, 81
*The Star*, 71
*State Authority* (Hill), 146
State Coal Mines, 116
State Hydroelectric Department (SHD), 109, 111. *See also* New Zealand Electricity Department (NZED)
Statute of Westminster Adoption Act 1947, 110
Stohler, Ann, 22
subject-colonialism, 10
Suppression of Rebellion Act 1863, 114
systematic colonization, 12, 13, 41

Taheke 8, 165
Tainui (confederation of Māori iwi), 18, 104–5, 112, 114–15, 120–32, 135. *See also* Huntly Power Station
Tainui Māori Trust Board, 129–30
*Tangata Whenua* (Anderson, Binney, and Harris), 20
taonga, 153, 155, 157
Tasman, Abel, 40
Tauhara North No. 2, 165
Taupiri Coal Mining Company, 115
Taupo Totara Timber Company, 152–53
Taupō Volcanic Zone (TVZ), 142, 153–55
Taylor, Thomas Edward, 88–91, 93
Te Ahi O Maui Plant (25 MW), 165
Te Arawa, 141–42
technology: as an analytic tool, 10–12; modernity and, 38
Te Ika-a-Māui, 141, 146–47; Parihaka Māori settlement, 147
Te Kupenga Charitable Trust, 150
*The Telegraphic Journal and Electrical Review*, 26
telegraph/telegraphic communication, 17, 25, 79, 85, 102, 191n51
Te Pāti Māori, 137
Te Reo Māori, 125, 164, 172
Te Ture Whenua Māori Act, 153

Teviot River, 172
Te Waipounamu, 146
*Thames Adviser*, 70
*Thames Star*, 27
Thermal Spring Districts Act of 1881, 142
Thompson, Sylvanus, 49–50
Thomson, William "Lord Kelvin," 9, 48, 96
Tikitere Geothermal Company, 165
Tillis, Emma, 51
tourist industry, 50, 52, 96, 147, 150. *See also* geothermal tourism industry
Town and Country Planning Act 1953, 126–28
Town and Country Planning Committee, 130
Transpower, 162
Treaty No. 9, 100
Treaty of Waitangi, 21, 43, 72, 73, 84, 142; breaches of, 151; government spurning, 143; manipulative land acquisition, 114; Te Ture Whenua Māori Act and, 153
Treble Government, 81
Trilemma rating index, 164–65. *See also* World Energy Council
Tūaropaki Development Scheme, 153
Tūaropaki E. (land), 156
Tūaropaki Land Development Scheme, 153
Tūaropaki Power Company, 18–19, 153–66; dairy processing plant and, 160; mission, 153; profits and energy, 158. *See also* Tūaropaki Trust
Tūaropaki Trust, 18, 140, 146, 152–53, 170; climate-controlled glasshouse, 159; dairy processing plant, 160; decision-making process, 163; developing Mōkai geothermal fields, 155; goal, mission, and principles, 158, 162; Gourmet Mokai, 159–60; Māori agricultural interests, 152; partnerships, 159–60; sustainability, 158; Waitangi Tribunal, 153
*Tupuna Awa* (Muru-Lanning), 146
Turrettini, Théodore, 90
Tuuta, Dion, 146
TVZ. *See* Taupō Volcanic Zone

Ulrich, George Henry Frederick, 55–56
United States: Niagara Falls, 49, 89; water regulations, 99–100
University of Otago, 69
urbanization, 107–8

Vogel, Julius, 16, 28, 32, 39, 44–46, 50, 57, 83–85, 107, 115

Waahi O Marae Committee, 124
Waahi Pā Marae, 122–24
Waihi Gold Mining Company, 113
Waikato, 112, 114. *See also* Tainui (confederation of Māori iwi)
Waikato Rauputu Claims Settlement Act (1995), 204n56
Waikato River, 25, 112–16, 118, 122, 127–29, 131, 132, 150; as living ancestor of Tainui, 114
Waikato River Valley Authority, 127, 128
Waikato Wars, 115, 122
Waimakariri River, 87–93
Wairakei Power Station, 149

Waitaki Schemes, 109
Waitangi Tribunal, 18, 129–30, 141, 151, 153, 164, 204n56; Land Grants, 114
Waititi, Rawiri, 137
Waiwera Geothermal Fields, 142
Wakefield, Edward Gibbon, 12, 13, 78
Walding, J. A., 118
Wanhalla, Angela, 163
Ward, Joseph, 76
Water and Soil Conservation Act 1967, 127, 128
Water-Power Act 1903, 92, 100, 113, 131
water rights, 99–100; Huntly legislation, 104–5; Māori and, 100; NZED, 127, 129; Tainui and, 123, 128–30
Watt, Hugh, 111, 119
*Webs of Empire* (Ballantyne), 20
*Weekly Press*, 27
Whātāpaka Marae, 151
*Wi Parata v. Bishop of Wellington*, 143
working-class migrants, 29
World Energy Council, 164–65. *See also* Trilemma rating index
World War I, 105, 107
World War II, 105, 109

Zero Carbon Act, 3
zero-carbon hydrogen production, 161

www.ingramcontent.com/pod-product-compliance
Lightning Source LLC
LaVergne TN
LVHW090023251026
850939LV00002B/13